Galaxies, Nebulae and Star Clusters:
A Photographic Tour

Anthony LaConte

Magical Moose Publishing

Galaxies, Nebulae and Star Clusters:

A Photographic Tour

Published by **Magical Moose Publishing**

First Edition

ISBN 979-8-9971974-9-0

Photographs by Anthony LaConte

Book design by Anthony LaConte

Cover design by Anthony LaConte

Printed in the United States of America

Special Thank You,

To my wife Carole,

My forever love and best friend.

For all her support in all that I do.

For encouraging me for the past 6 years to create a book.

For her patience and support all the nights I was outside imaging.

To my cousin Larrie LaConte,

for her outstanding editing and proofing during the creation of this book.

For sharing family memories of our childhood while we did the editing.

To all my fellow amateur astronomers and stargazers

who shared their observing tips and friendship over the past 50 years.

Introduction

We often start out "stargazing" as a fun hobby, enjoying nature and finding shapes or groups of stars in the night sky. As our curiosity grows, we want to know more about those slowly drifting seasonal patterns—the constellations. We notice small, fuzzy areas of light and wonder what they could be. We progress from looking at the sky with just our eyes to getting more detailed views with binoculars, and then to exploring the universe with a telescope. Some of us even go on to study the universe as amateur astronomers or professional scientists.

Ancient people had a major advantage over today's stargazers: no light pollution. The night sky was black as coal, with only the glow of the Milky Way breaking the darkness. Constellations were easy to identify and revealed far more than a few bright stars. The "Seven Sisters," also known as the Pleiades, formed a brilliant cluster whose individual stars could be seen and counted.

Today's stargazers and amateur astronomers are constantly fighting the glow from big-city lights, suburban shopping areas, and unshielded residential street lighting. This light scatters into the night sky and hides what we so dearly want to see. The amateur astronomer has many accessories in their observing toolbox: binoculars for scanning large areas of the sky; a telescope for lunar and planetary viewing; another for exploring galaxies, nebulae, and star clusters; and cameras for capturing the wondrous sights the universe has long hidden from those of us who inhabit Earth.

Twenty-first-century technology now allows amateur astronomers to capture these celestial wonders in stunning color and extraordinary detail with sensitive, specialized astronomy cameras. But the task is complicated by light pollution. A simple solution is to travel somewhere away from major sources of artificial light—away from large cities, towns, and densely populated suburban areas. Basically, head for the mountains, forests, or secluded desert locations. While this can greatly enhance both viewing and photography, it is not always possible. For some amateur astronomers, the challenges of light pollution can even discourage them from attempting astrophotography.

All the images in this book were taken by the author, from the author's backyard, less than 10 miles from downtown Phoenix, Arizona. This is a Bortle 8 zone (city-suburban sky / naked-eye limiting magnitude 4-5 / Milky Way not visible / significant sky glow toward the Phoenix metro area). These images demonstrate what can be achieved despite the challenges of a heavily light-polluted sky.

The goal of this book is to inspire both young and old to explore and photograph the night sky—even from their own backyard.

See what is possible!!

Book Format

The core of the book is divided into four image category sections: Galaxies; Nebulae; Star Clusters; and the Solar System. Each image is labeled with an astronomical reference designation and a common name if one exists; the specific type of object it is; and the approximate distance from the Earth. It should be noted that I used the word "approximate" for the distances. Distances to astronomical objects are determined using a variety of measurement techniques, and different sources may state different distances. These "estimates" can be updated as new observations and improved measurement techniques are employed. If you do a web search on an object, you will find the distances stated may vary depending on the source. That does not necessarily mean one source is correct and another is wrong; rather, the distances are calculated estimates based on the observations and measurement techniques available at the time.

Appendix 1 is the list of images grouped by constellation. This provides a quick lookup to see which constellation each image is located in and on which page the image can be found.

Appendix 2 is the imaging data for every image in the book. A table of common equipment used for all the images is at the beginning of the appendix. The imaging data for each image is presented in page number order. For every image, I have included the color palette, exposures used, telescope, reducer/flattener, imaging filters, and camera.

For **Messier Catalog** objects, the image captions spell out **"Messier."** In the appendices, the shortened form—***M*** followed by the catalog entry number—is used.

The initialism **"NGC"** refers to objects listed in the ***N***ew ***G***eneral ***C***atalog. NGC is followed by the catalog entry number.

The Imaging Equipment Used

Sky-Watcher 150 DX doublet APO refractor
Aperture: 150 mm
Focal Length: 1200 mm
F/ratio: 8
Sky-Watcher 0.77 Reducer/Flattener
ZWO EAF (Electronic Automatic Focuser)

Sky-Watcher 180 mm Maksutov-Cassegrain
Aperture: 180 mm
Focal Length: 2700 mm
F/ratio: 15

Sky-Watcher Esprit 80 ED Triplet APO refractor
Aperture: 80 mm
Focal Length: 400 mm
F/ratio: 5
Sky-Watcher Flattener
ZWO EAF (Electronic Automatic Focuser)

Stellarvue SV 60 ED APO Refractor (Guide scope)
Aperture: 60 mm
Focal Length: 330 mm
F/ratio: 5.5

Camera: ZWO ASI 2600MM – (Mono) Cooled
Resolution: 26 Megapixels (6248 x 4176)
Pixel Size: 3.76 µm

Camera: ZWO ASI 2600MC – (Color) Cooled
Resolution: 26 Megapixels (6248 x 4176)
Pixel Size: 3.76 µm

Camera: ZWO ASI 174MM mini (Guide Camera)
Resolution: 2.3 Megapixels (1936 x 1216)
Pixel Size: 5.86 µm

ZWO 7 position EFW (Electronic filter wheel)

Antlia Broadband Filters
Antlia V-Pro Luminance Filter
Antlia V-Pro Red Filter
Antlia V-Pro Green Filter
Antlia V-Pro Blue Filter

Antlia Narrowband Filters
Antlia Edge H-Alpha Filter (Hydrogen-alpha / 656 nanometers)
Antlia Edge OIII Filter (Oxygen III / 501 nanometers)
Antlia Edge SII Filter (Sulfur II / 672 nanometers)

Mount: Celestron CGX-L Computerized Equatorial Mount and Tripod
Pier Extender: Starizona custom 12-inch pier extender for the CGX-L mount / tripod

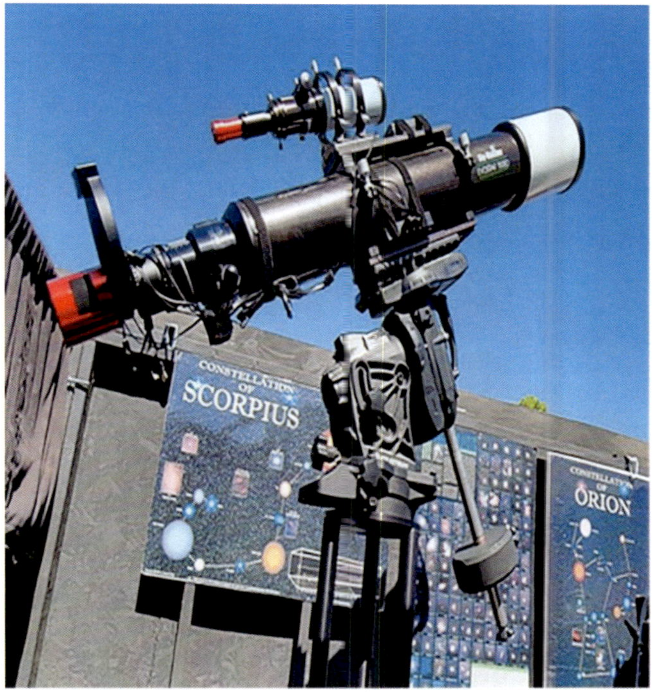

(A)

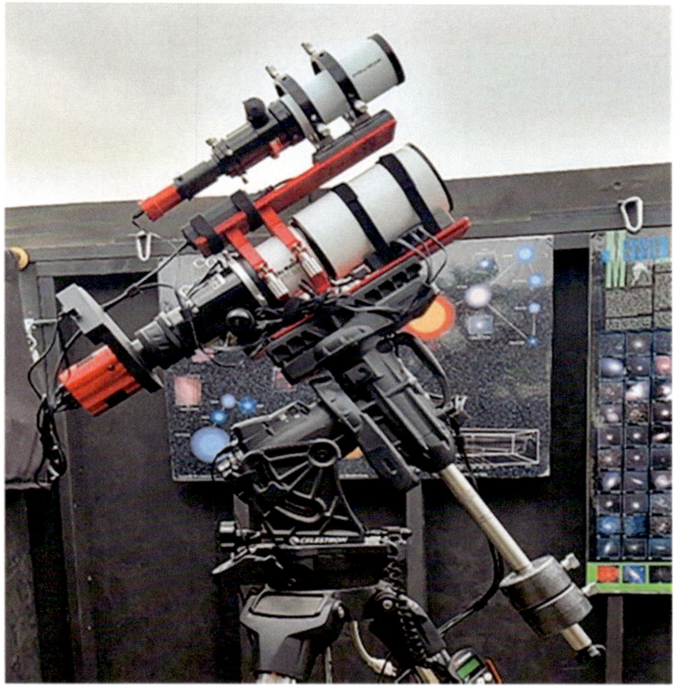

(B)

This is the equipment used to take the images in this book.

(A) 6 inch 150 DX (B) 80mm Esprit 80 ED (C) 7-position 2 inch filter wheel (D) ZWO 2600mm & 2600mc cameras & Antlia filters

(C)

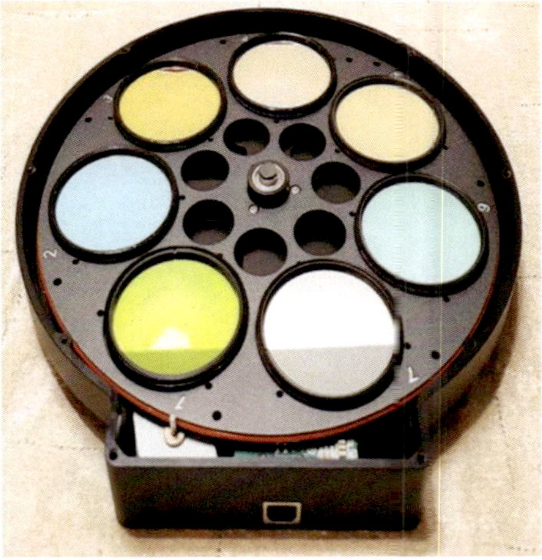

(D)

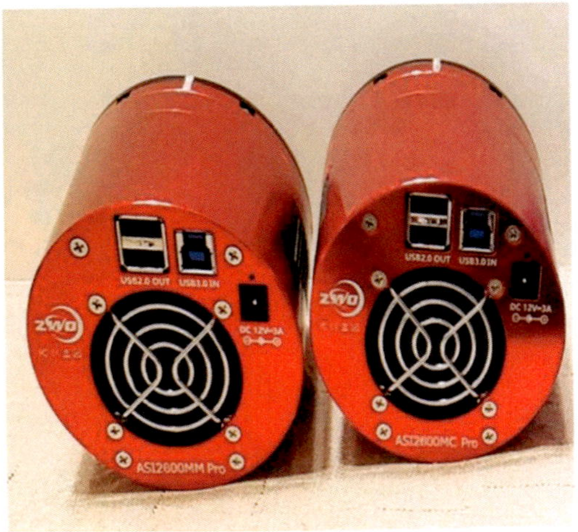

(D)

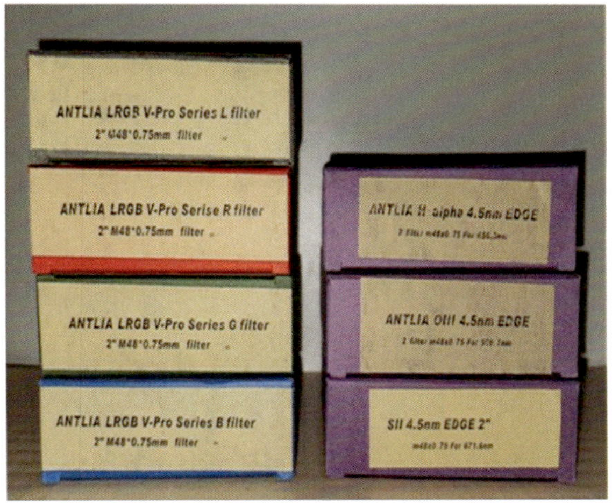

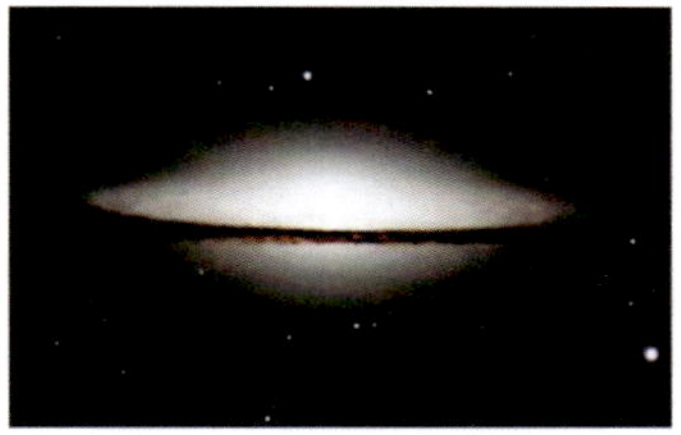

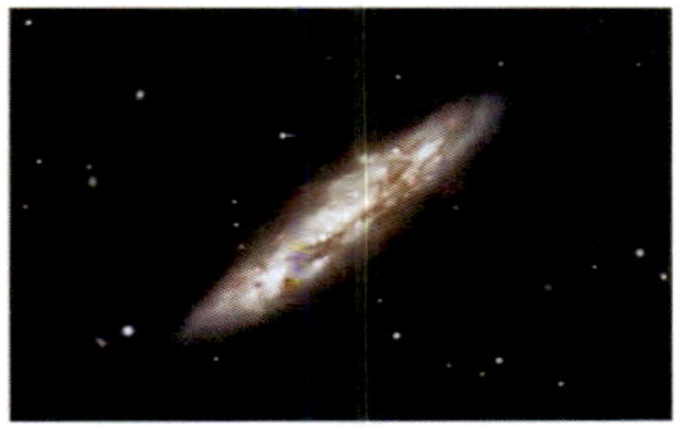

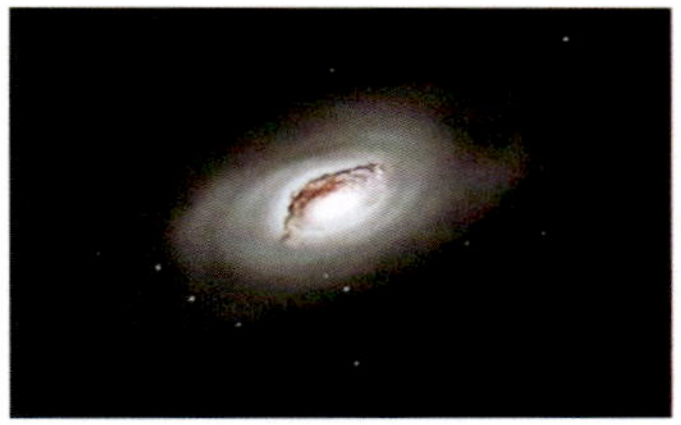

GALAXY IMAGES

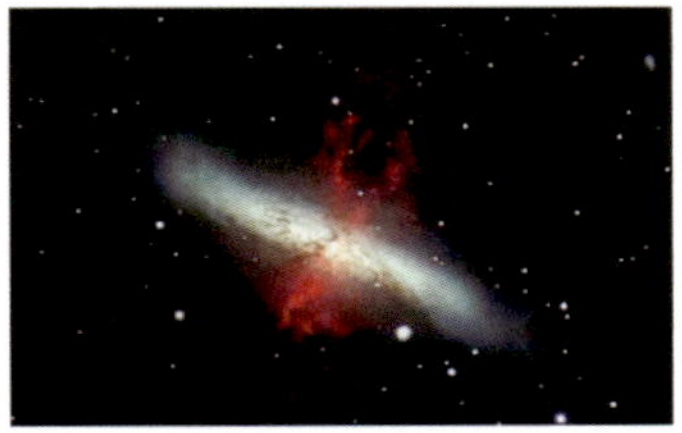

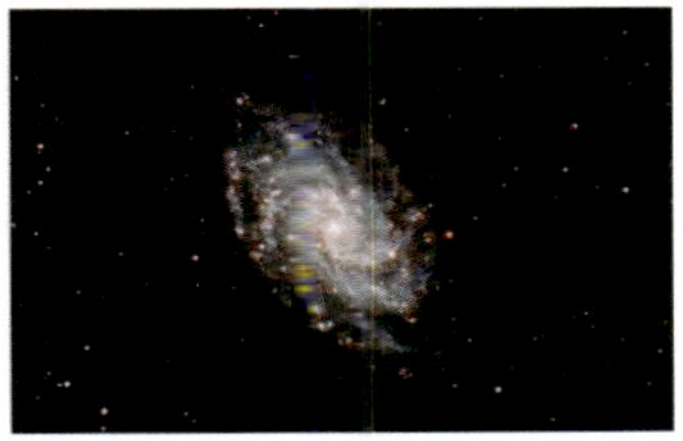

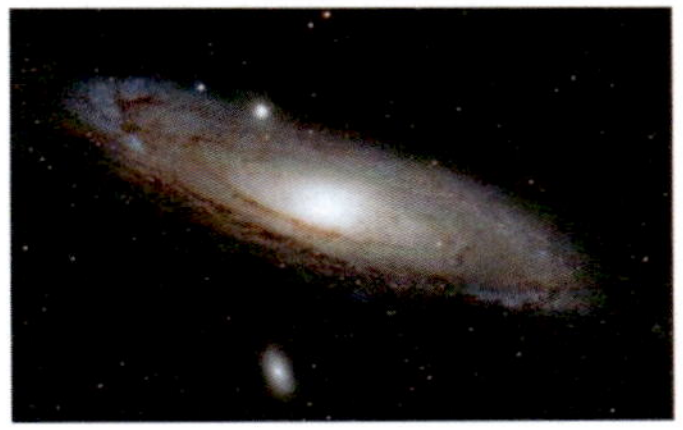

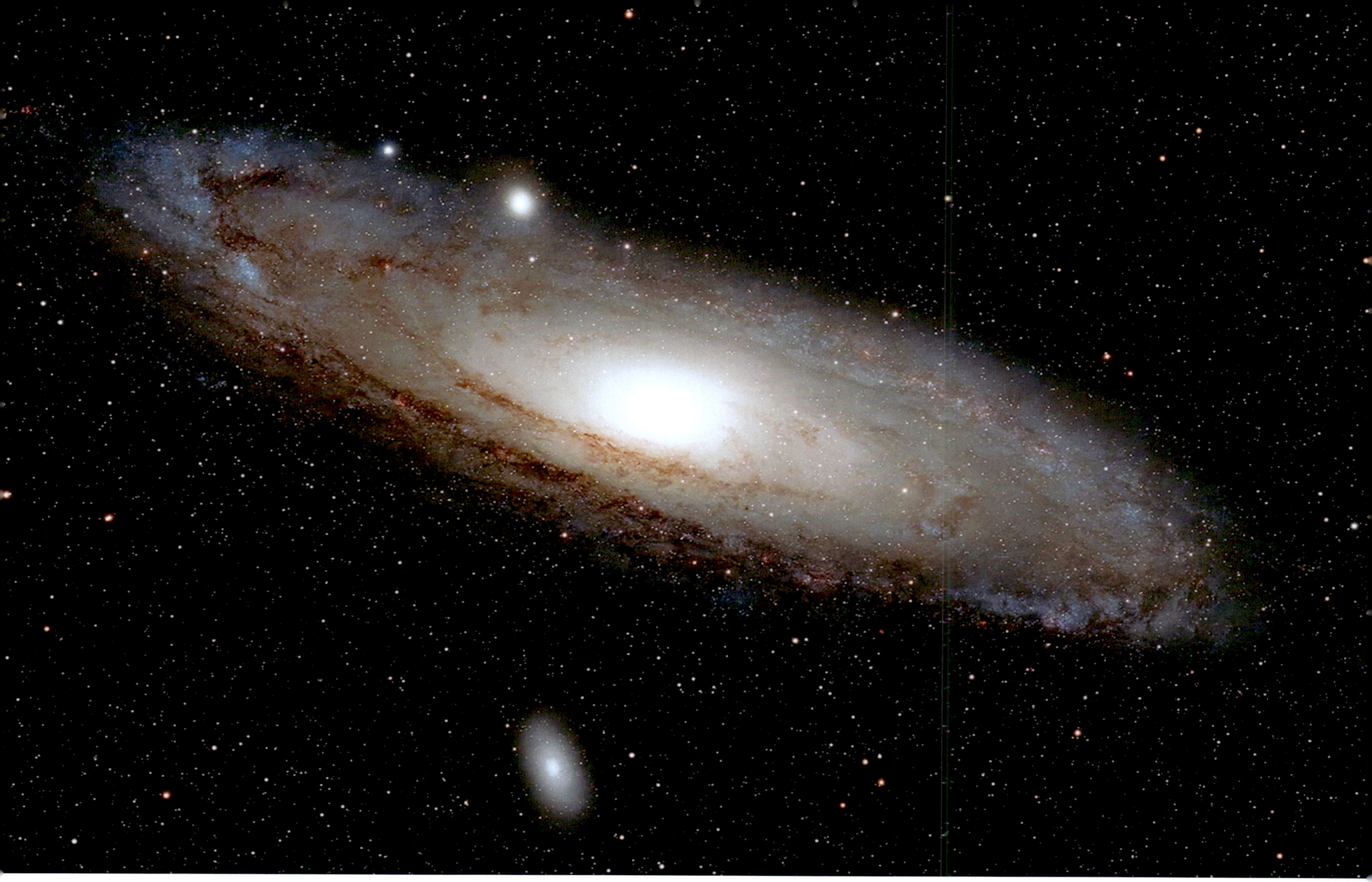

Messier 31 (The Andromeda Galaxy with Messier 32 (top) & Messier 110 (bottom))
A spiral galaxy in the constellation Andromeda
Approximate distance from Earth: 2.5 million light-years

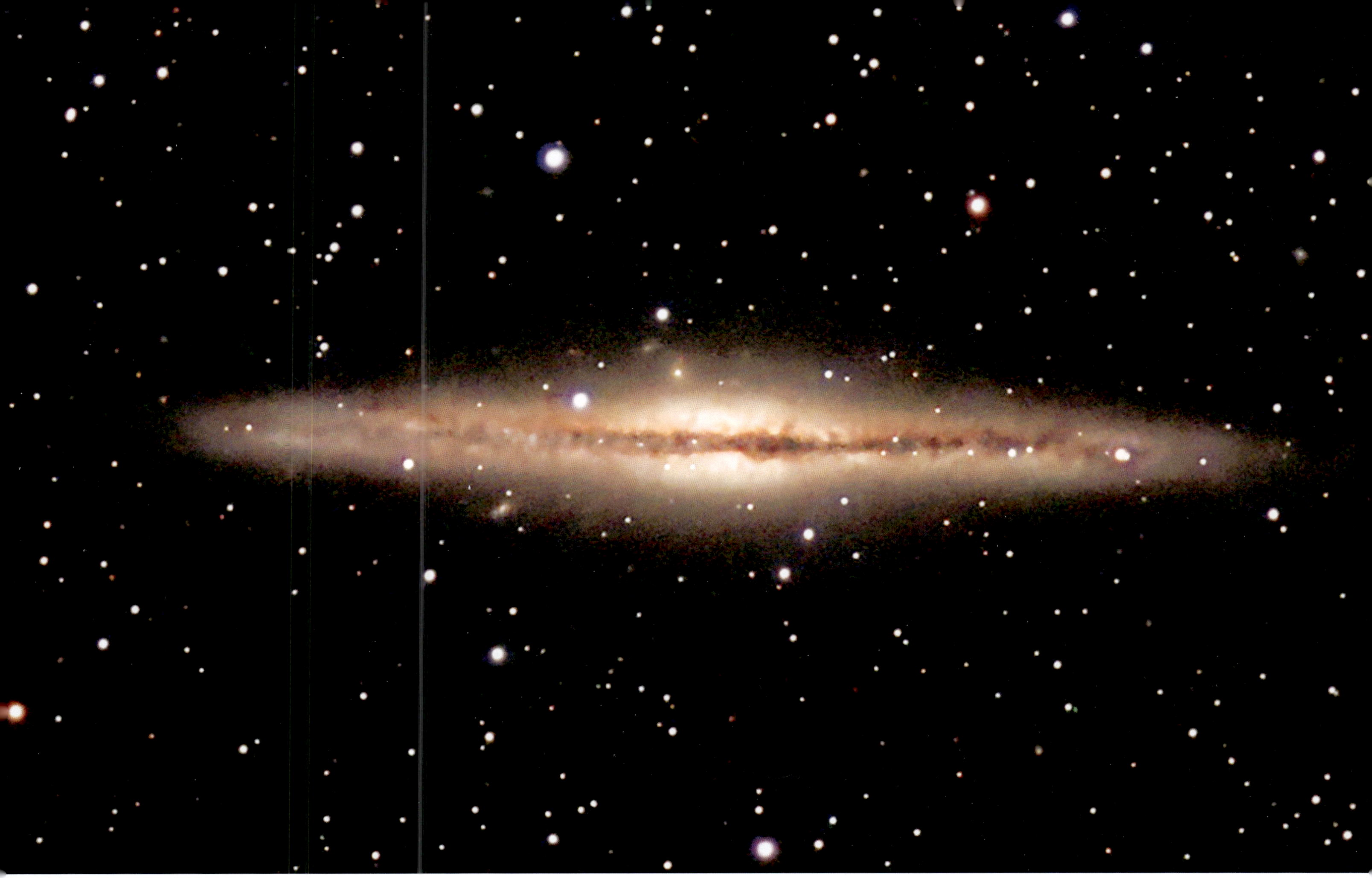

Outer limits Galaxy (NGC 891)
An edge-on spiral galaxy in the constellation Andromeda
Approximate distance from Earth: 31 Million light-years

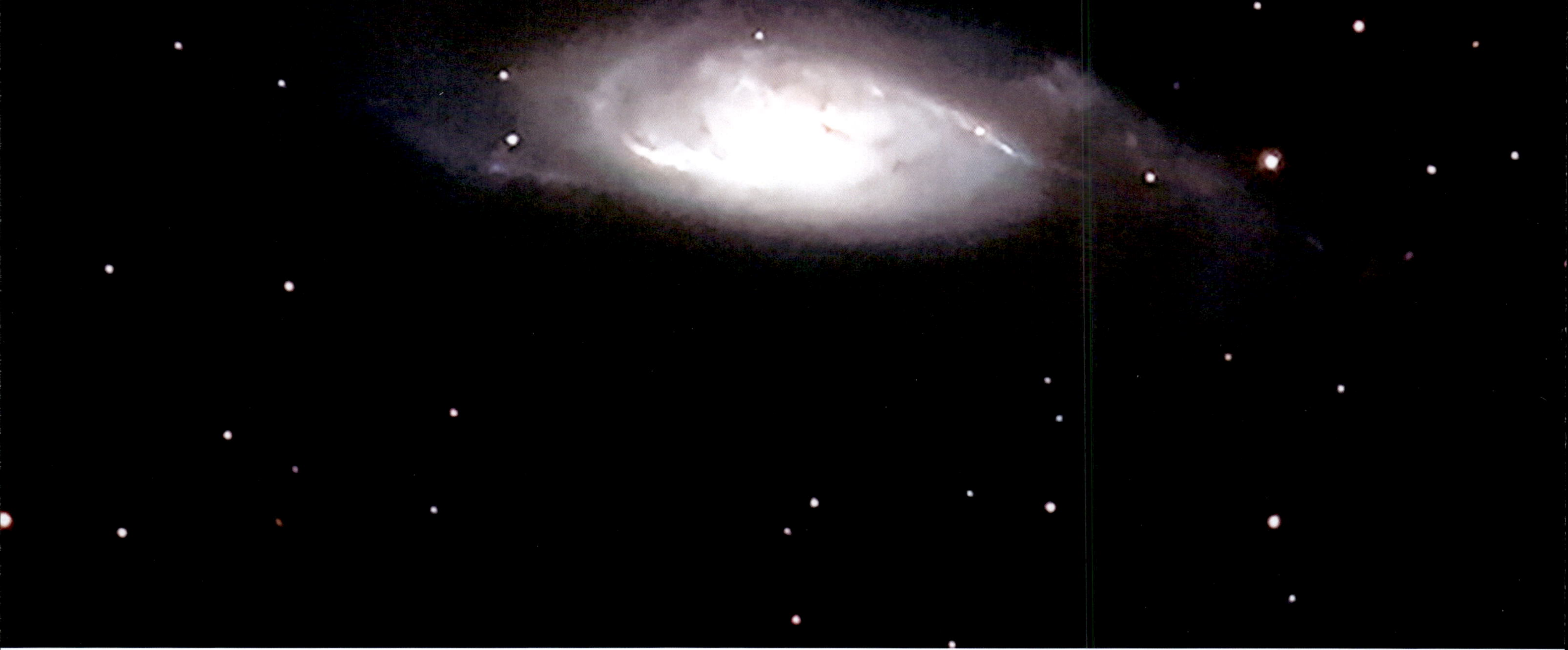

Messier 106
A spiral galaxy in the constellation Canes Venatici
Approximate distance from Earth: 24 Million light-years

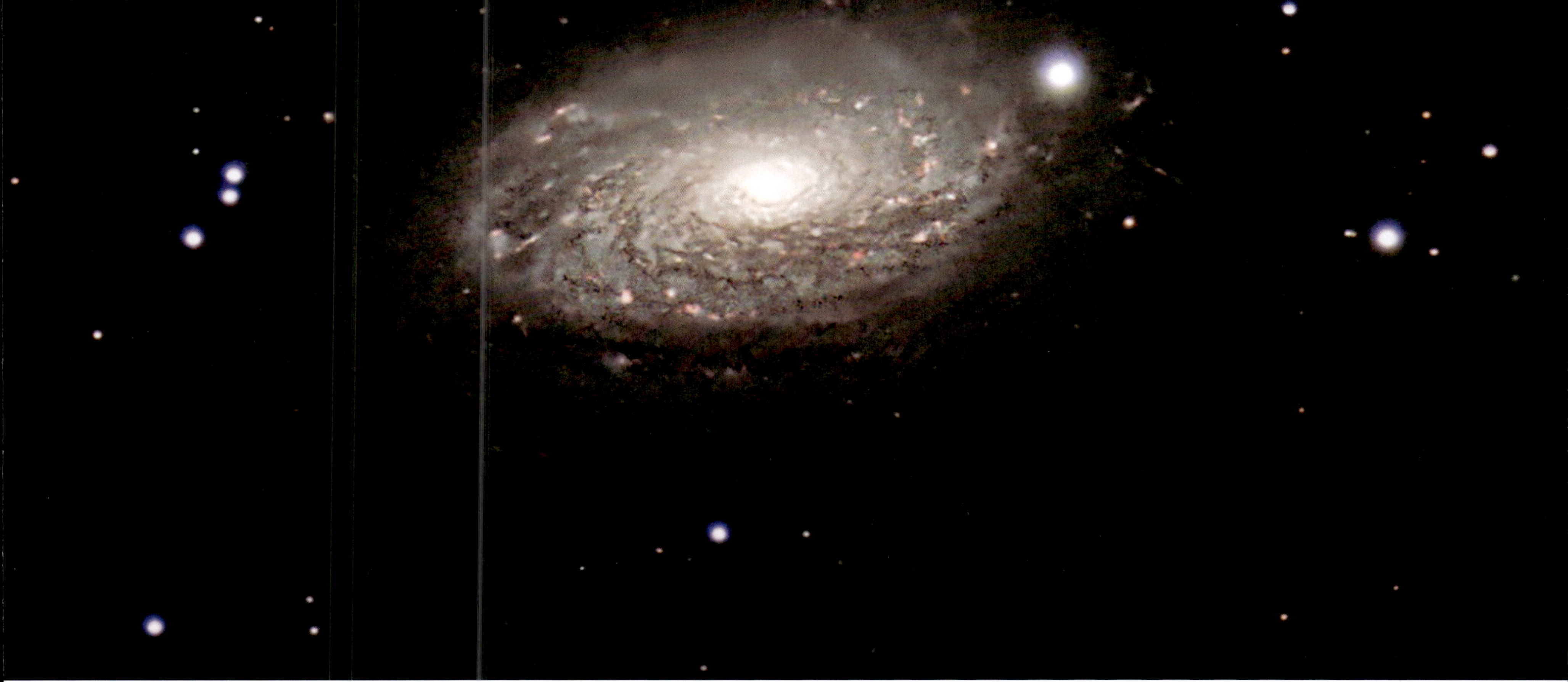

Messier 63 (The Sunflower Galaxy)
A spiral galaxy in the constellation Canes Venatici
Approximate distance from Earth: 25 Million light-years

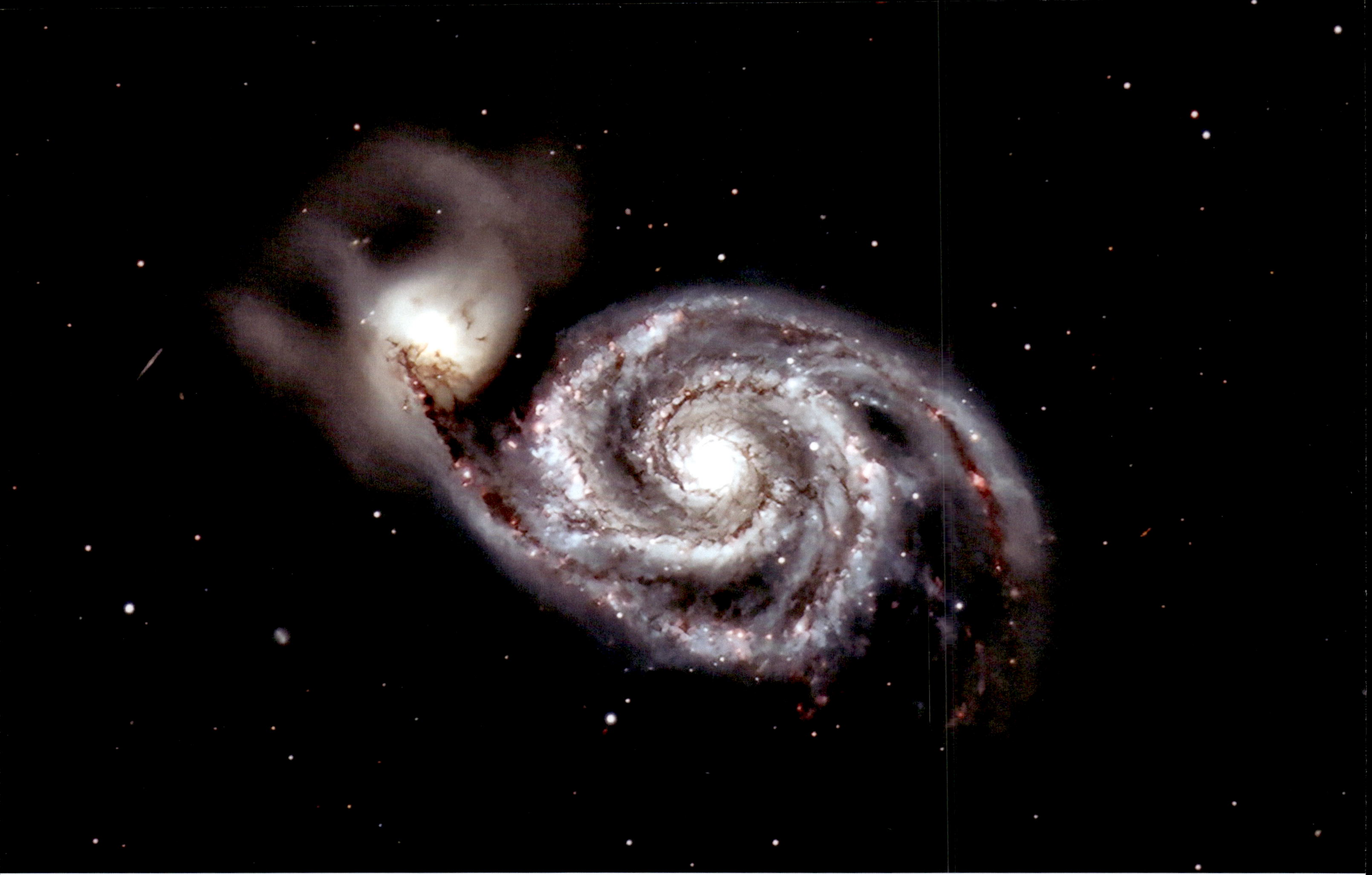

Messier 51 (The Whirlpool Galaxy interacting with the dwarf galaxy NGC 5195)
A spiral galaxy in the constellation Canes Venatici
Approximate distance from Earth: 26 Million light-years

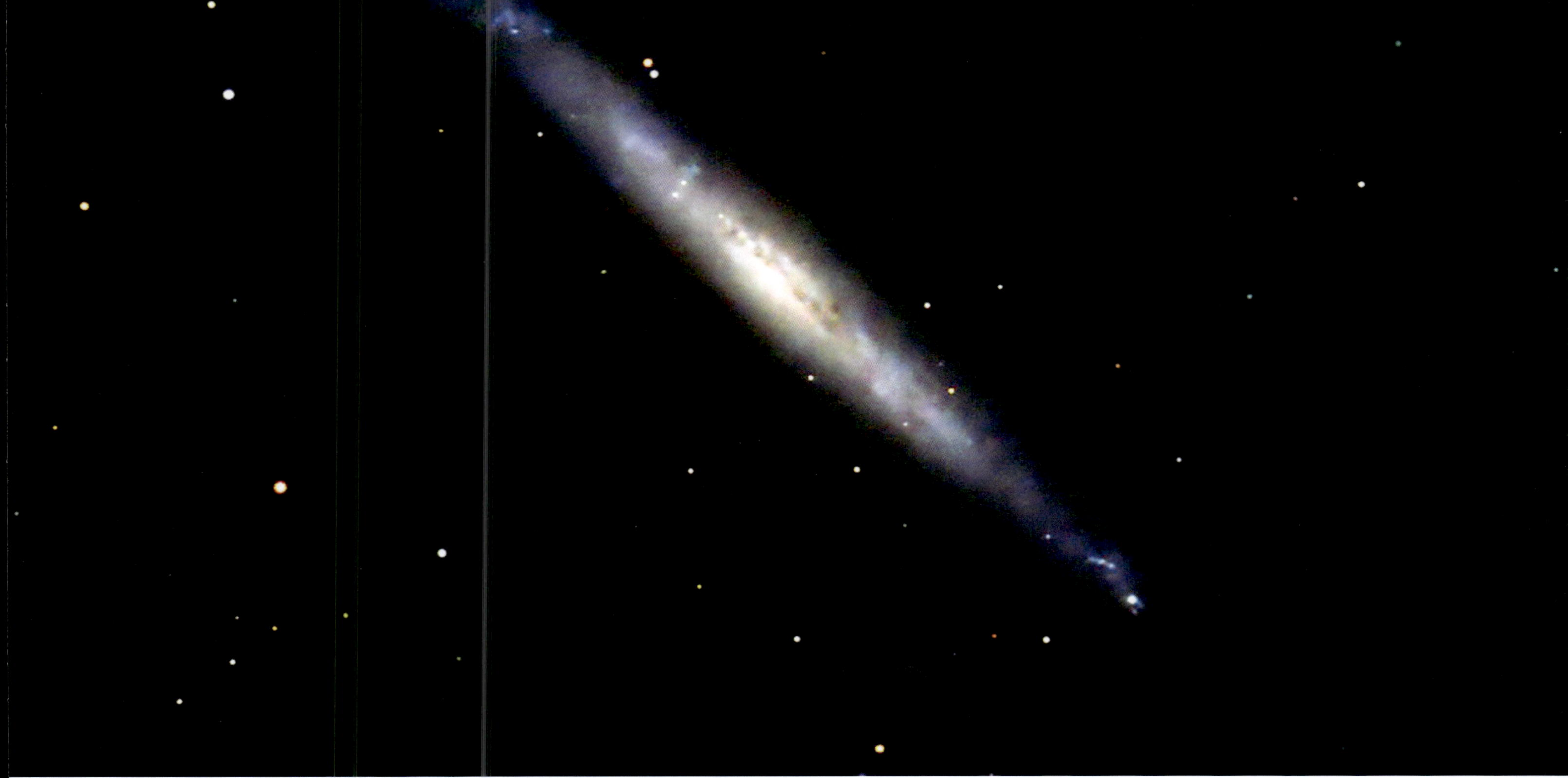

Silver Needle Galaxy (NGC 4244)
An edge-on spiral galaxy in the constellation Canes Venatici
Approximate distance from Earth: 10 Million light-years

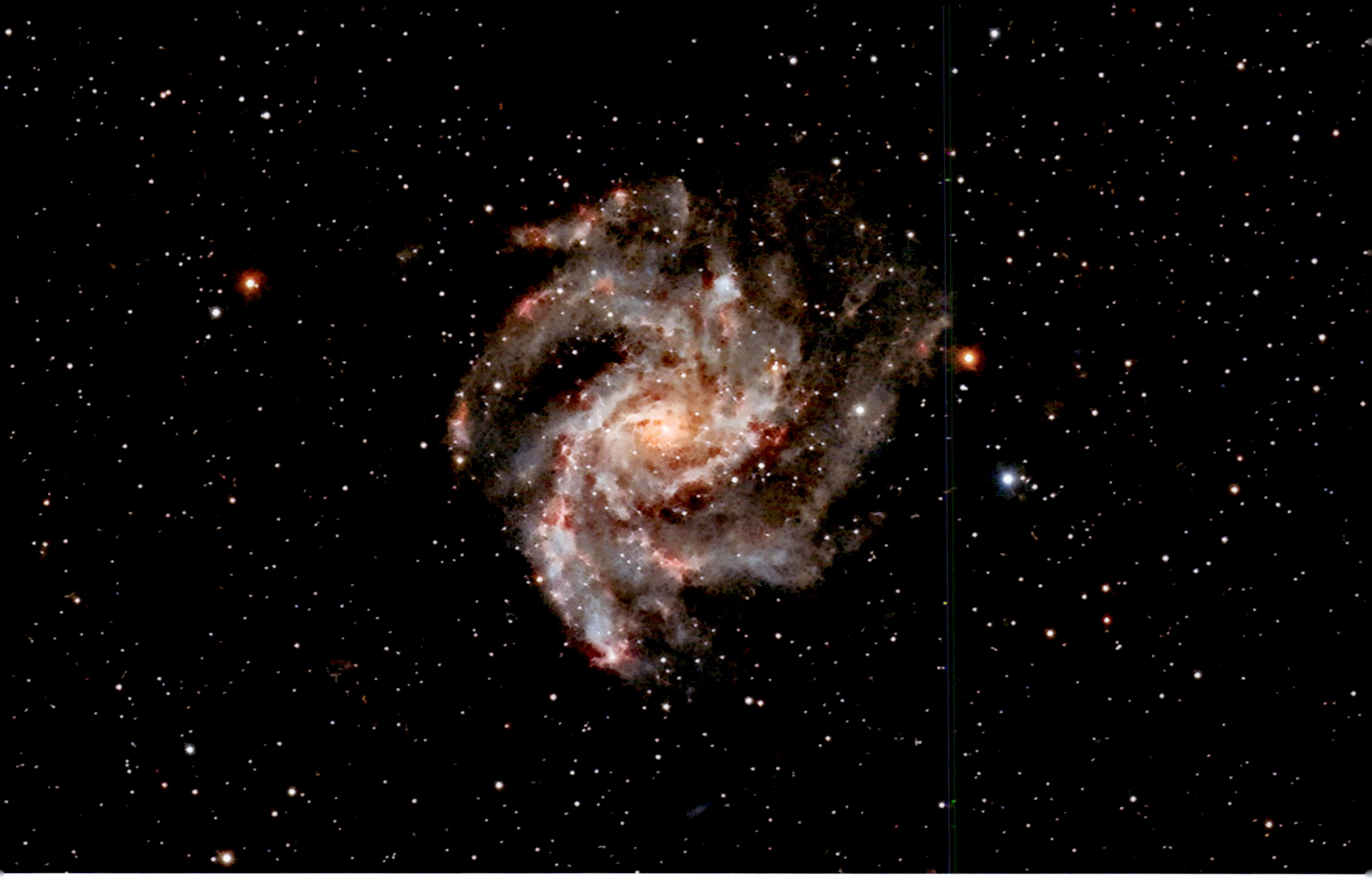

Fireworks Galaxy (NGC 6946)
A face-on spiral galaxy in the constellation Cepheus
Approximate distance from Earth: 25 Million light-years

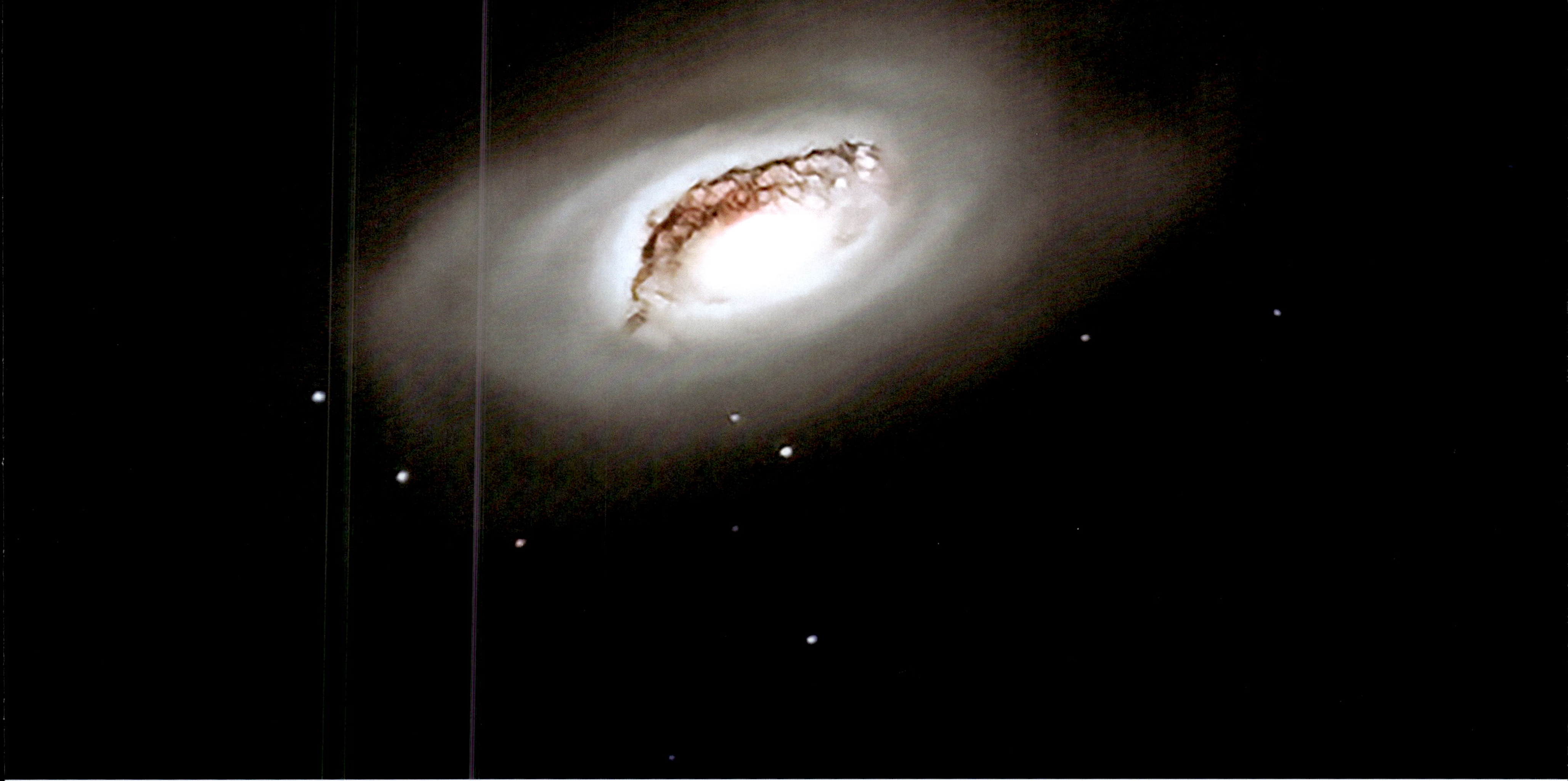

Messier 64 (The Black Eye Galaxy)
A spiral galaxy in the constellation Coma Berenices
Approximate distance from Earth: 24 Million light-years

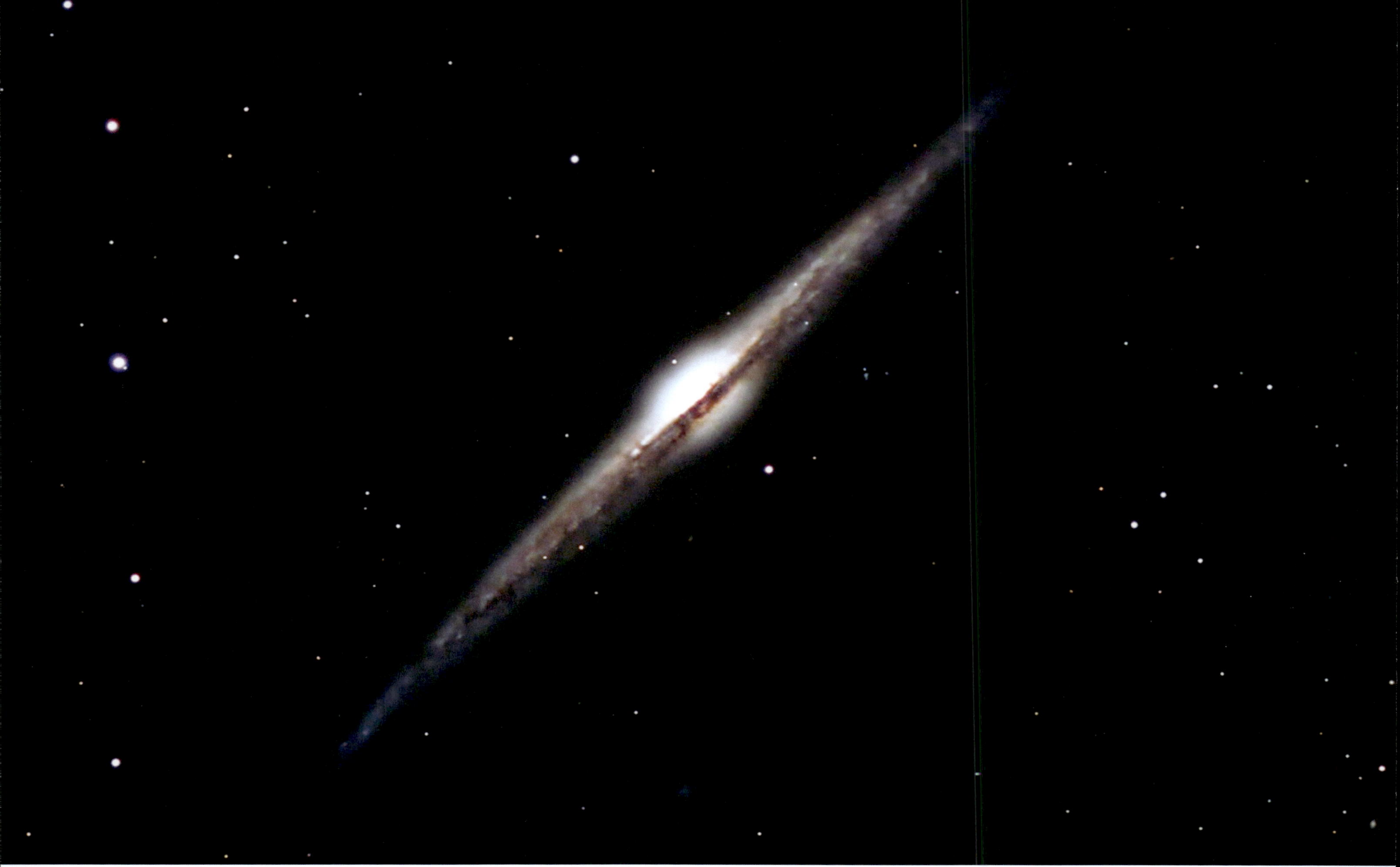

Needle Galaxy (NGC 4565)
An edge-on spiral galaxy in the constellation Coma Berenices
Approximate distance from Earth: 42 Million light-years

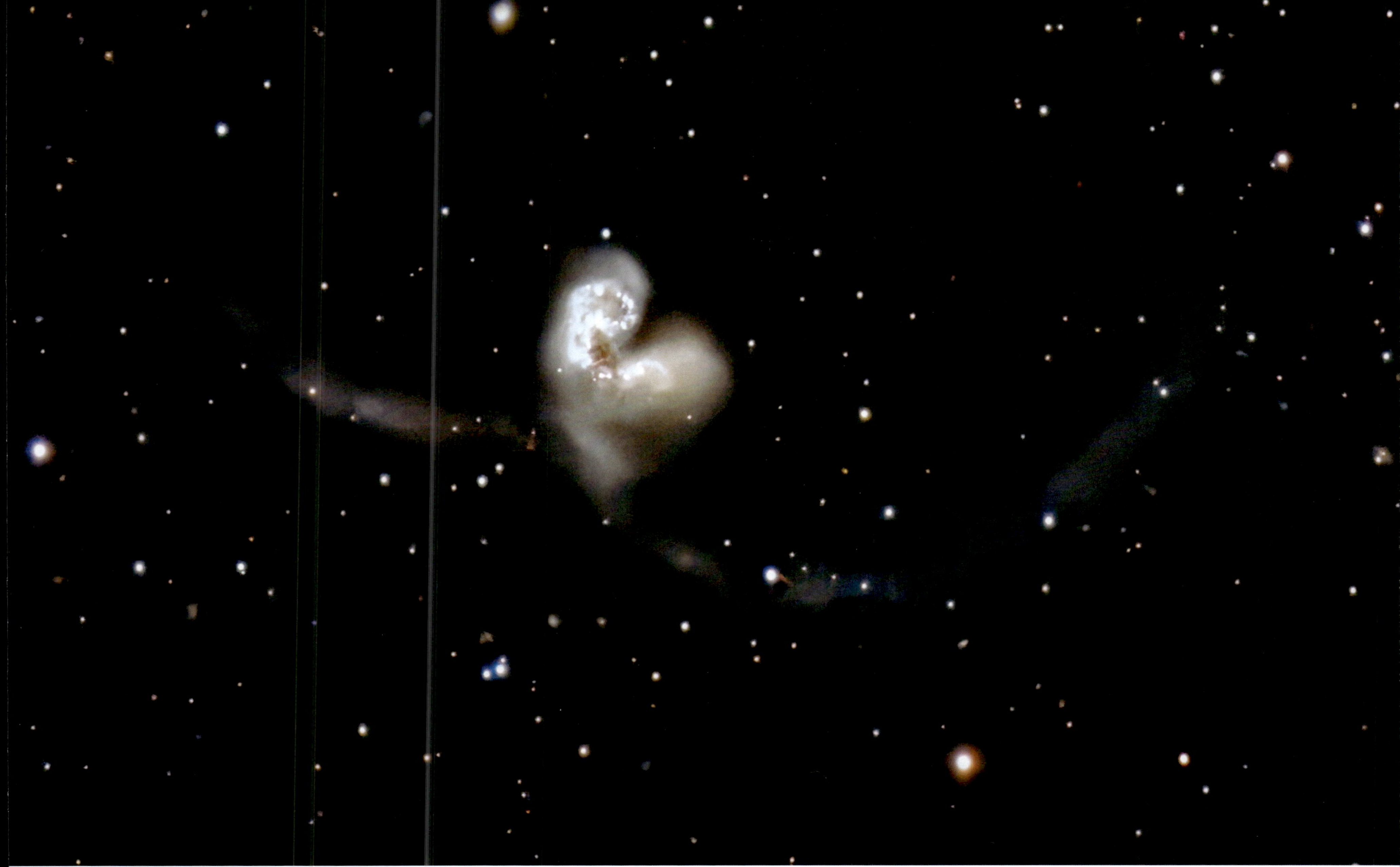

The Antennae Galaxies (NGC 4038 / NGC 4039)
Two galaxies colliding in the constellation Corvus
Approximate distance from Earth: 45 Million light-years

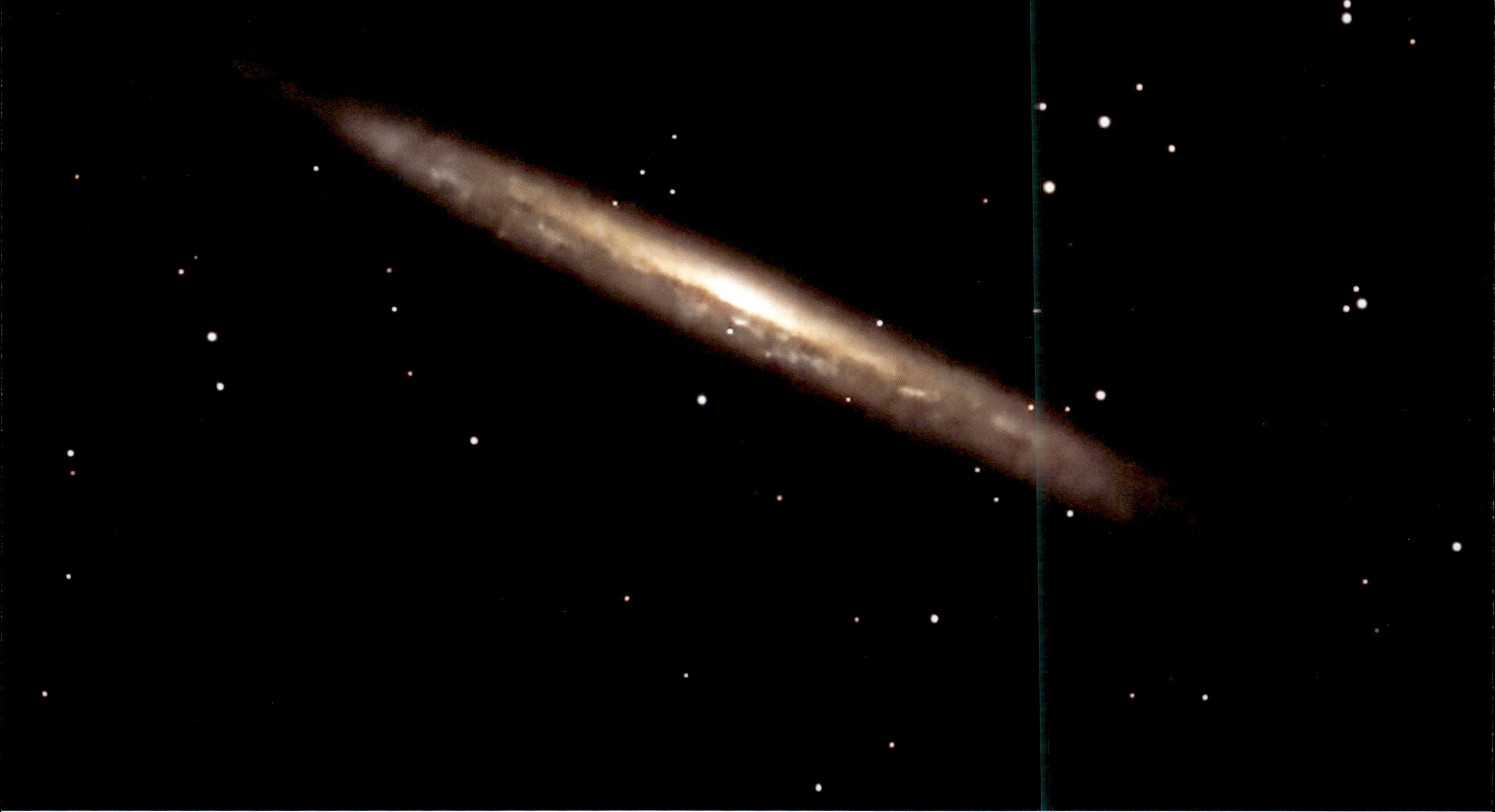

The Splinter Galaxy (NGC 5907)
An edge-on spiral galaxy in the constellation Draco
Approximate distance from Earth: 50 Million light-years

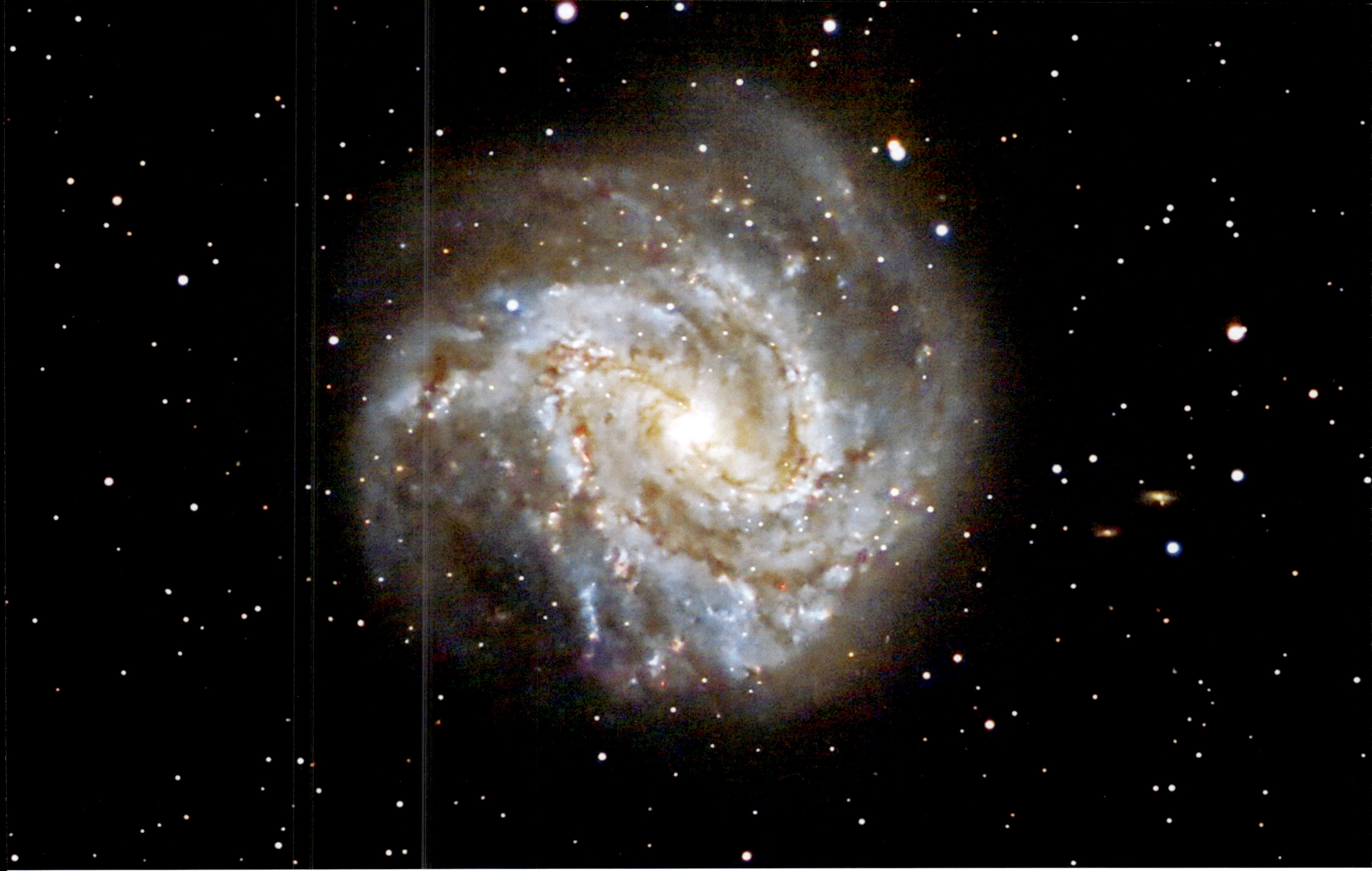

Messier 83 (The Southern Pinwheel)
A face-on barred spiral galaxy in the constellation Hydra
Approximate distance from Earth: 16 Million light-years

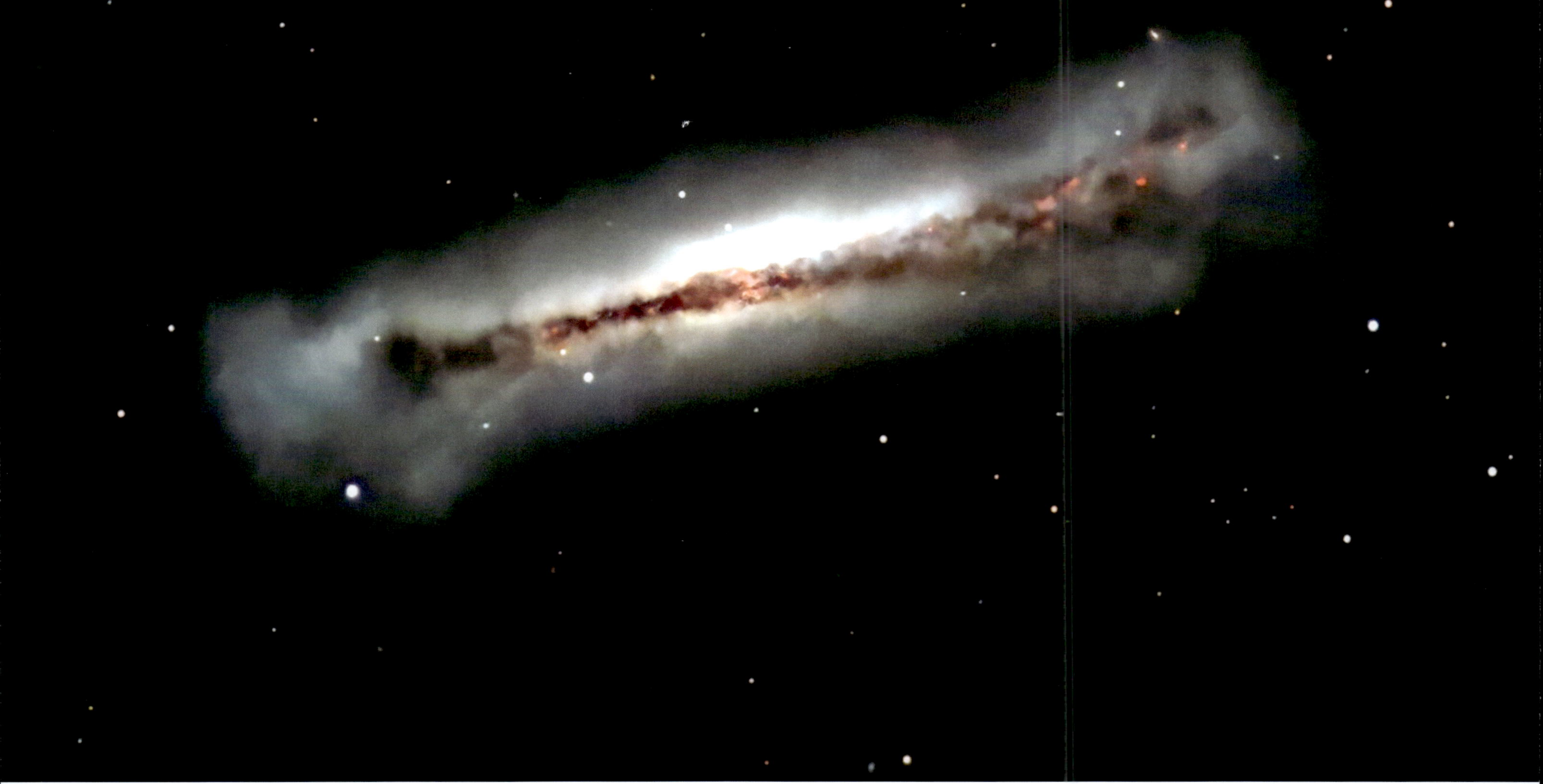

The Hamburger Galaxy (NGC 3628)
An edge-on spiral galaxy in the constellation Leo
Approximate distance from Earth: 35 Million light-years

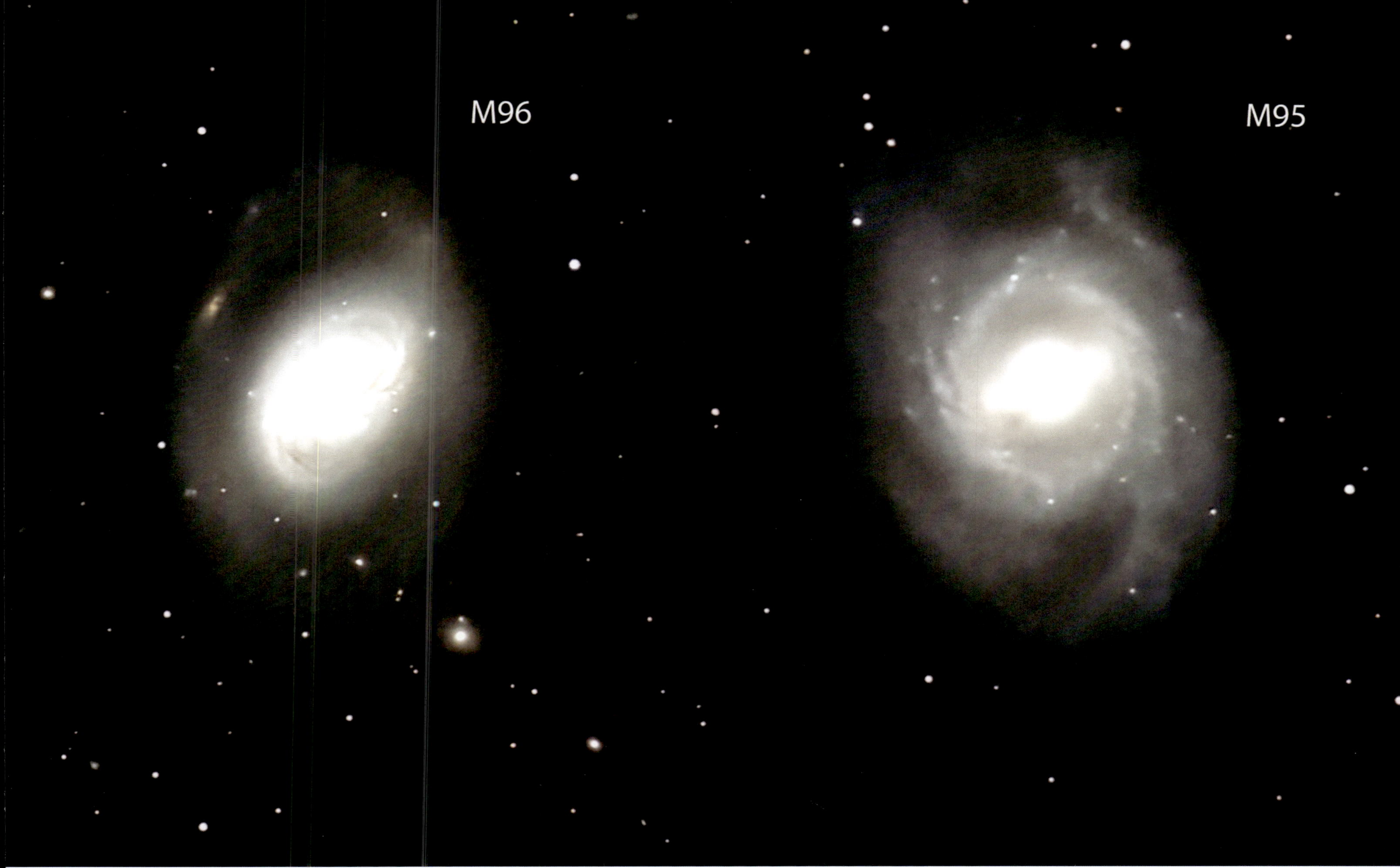

Messier 95 / Messier 96

Messier 95 and Messier 96 are both barred spiral galaxies in the constellation Leo

Approximate distance from Earth: 33 Million light-years

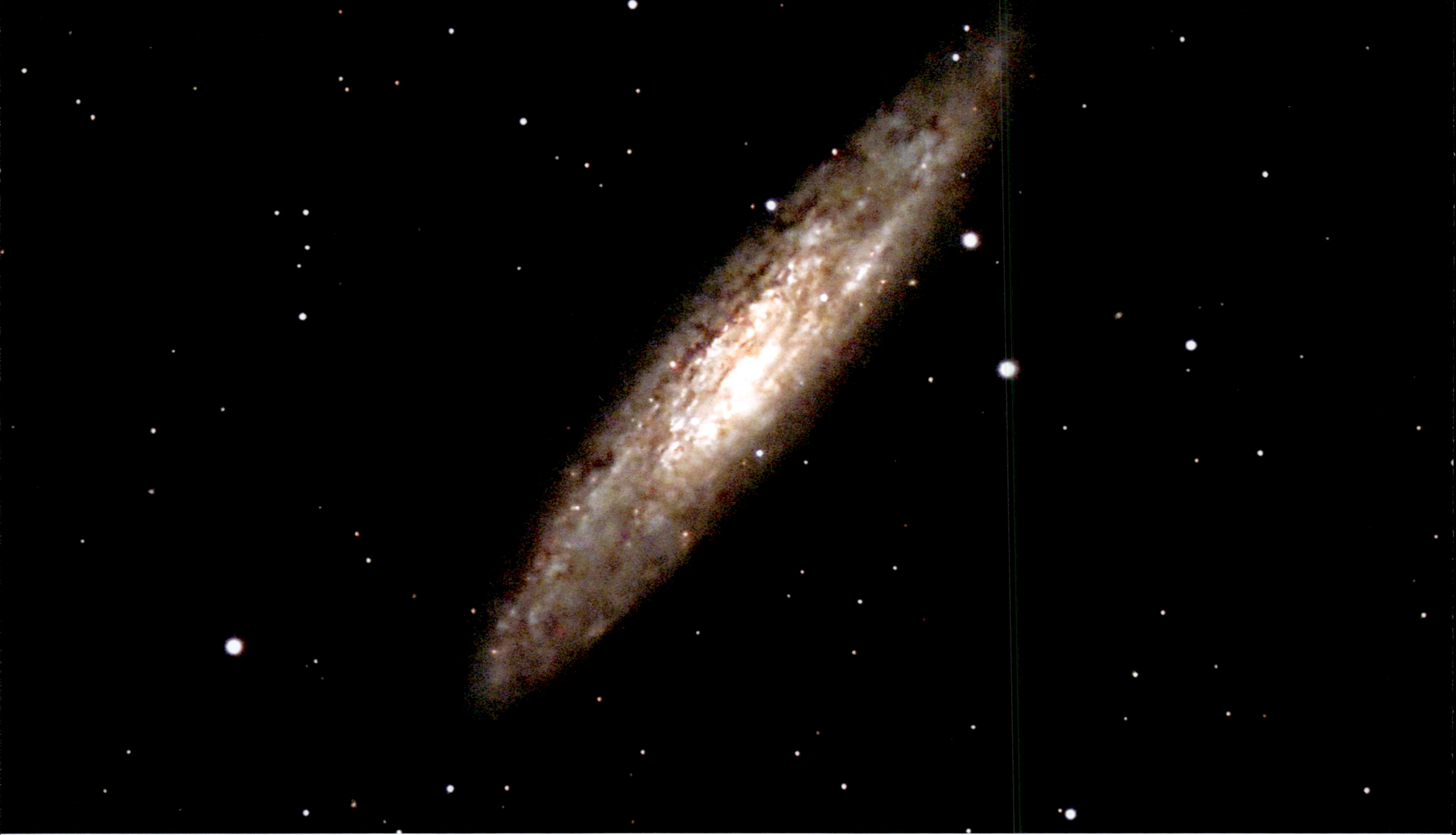

Sculptor / Silver Coin Galaxy (NGC253)
A spiral galaxy in the constellation Sculptor
Approximate distance from Earth: 10 Million light-years

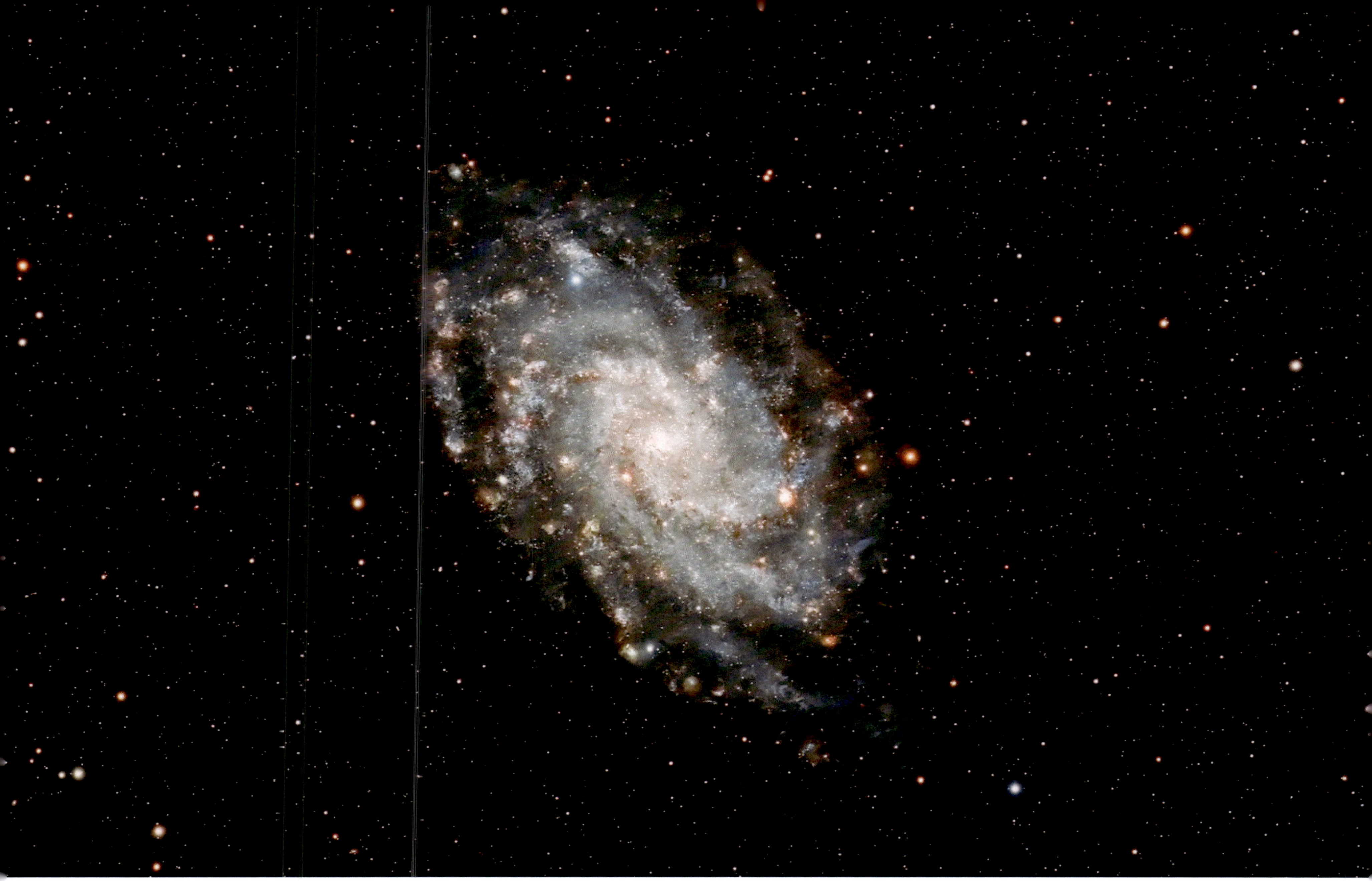

Messier 33 (Triangulum Galaxy)
A face-on spiral galaxy in the constellation Triangulum
Approximate distance from Earth: 2.7 Million light-years

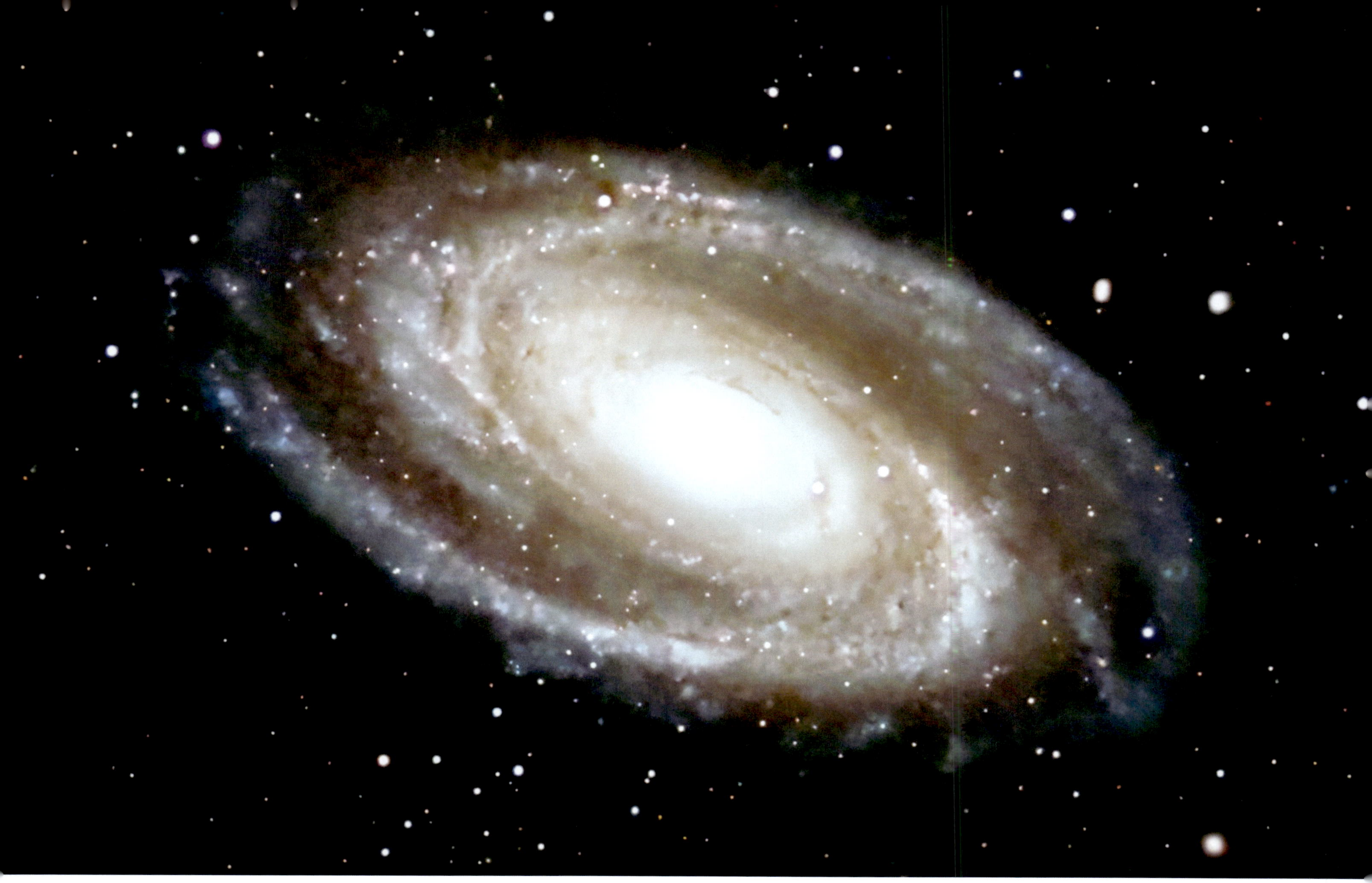

Messier 81 (Bode's Galaxy)
A face-on spiral galaxy in the constellation Ursa Major
Approximate distance from Earth: 11.6 Million light-years

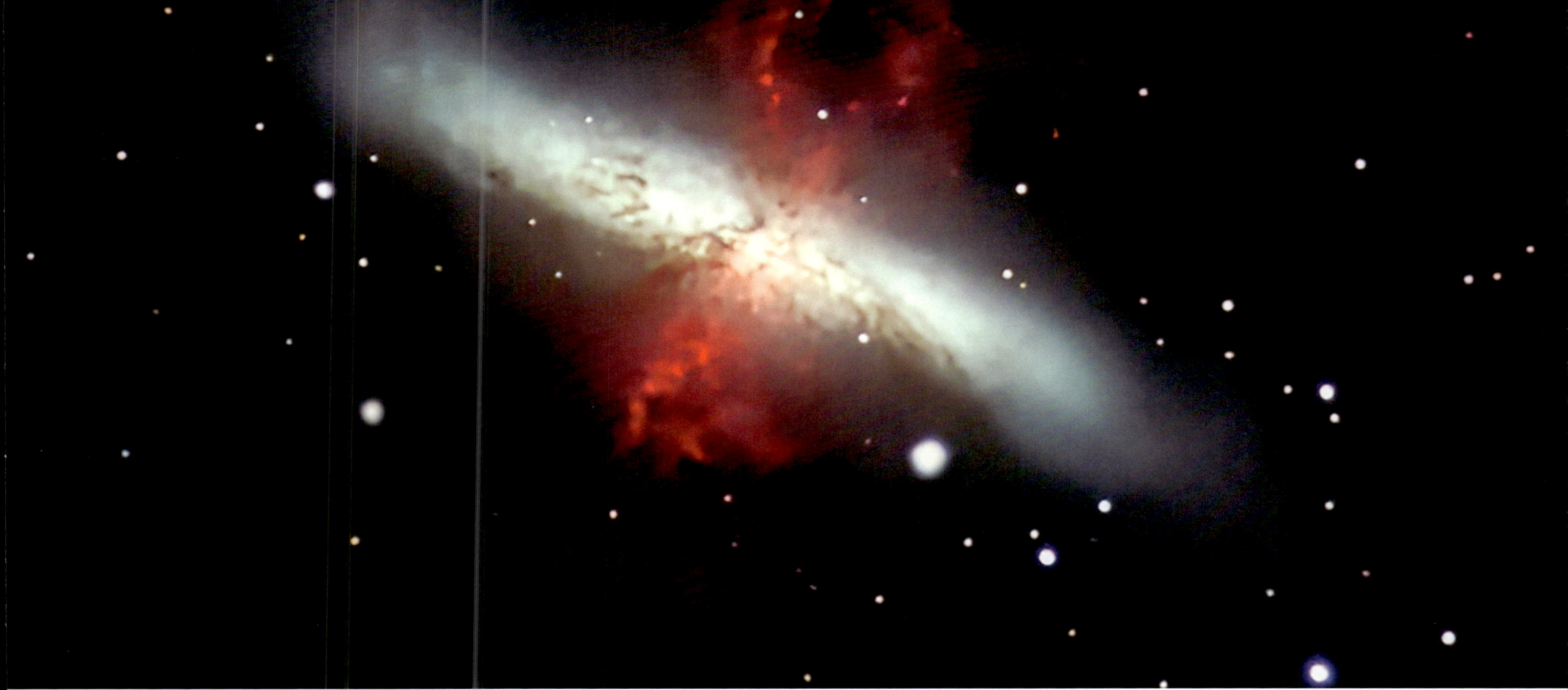

Messier 82 (The Cigar Galaxy)
An irregular galaxy in the constellation Ursa Major
Approximate distance from Earth: 12 Million light-years

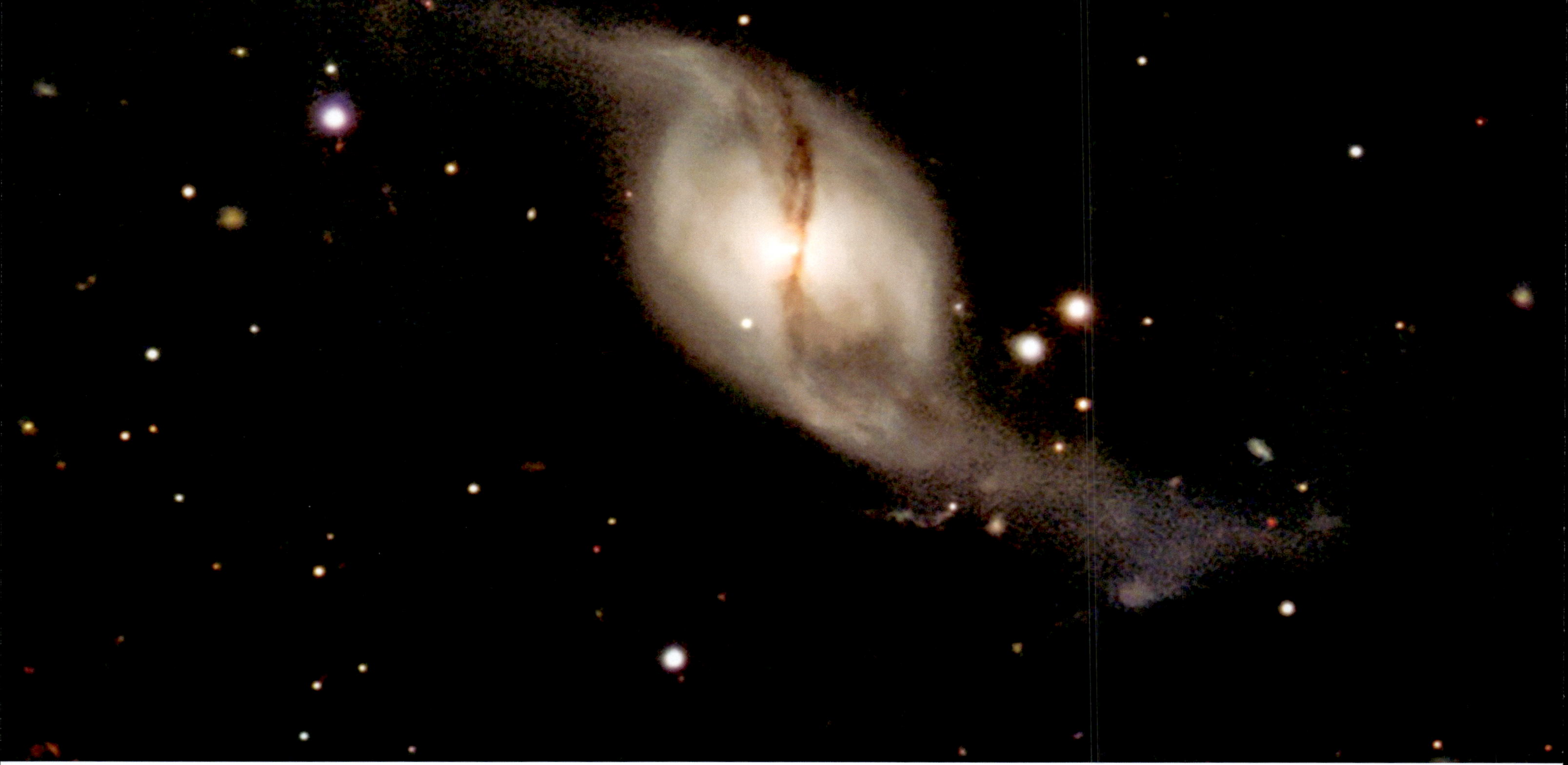

Arp 214 Galaxy (NGC 3718)
A peculiar barred galaxy in the constellation Ursa Major
Approximate distance from Earth: 56 Million light-years

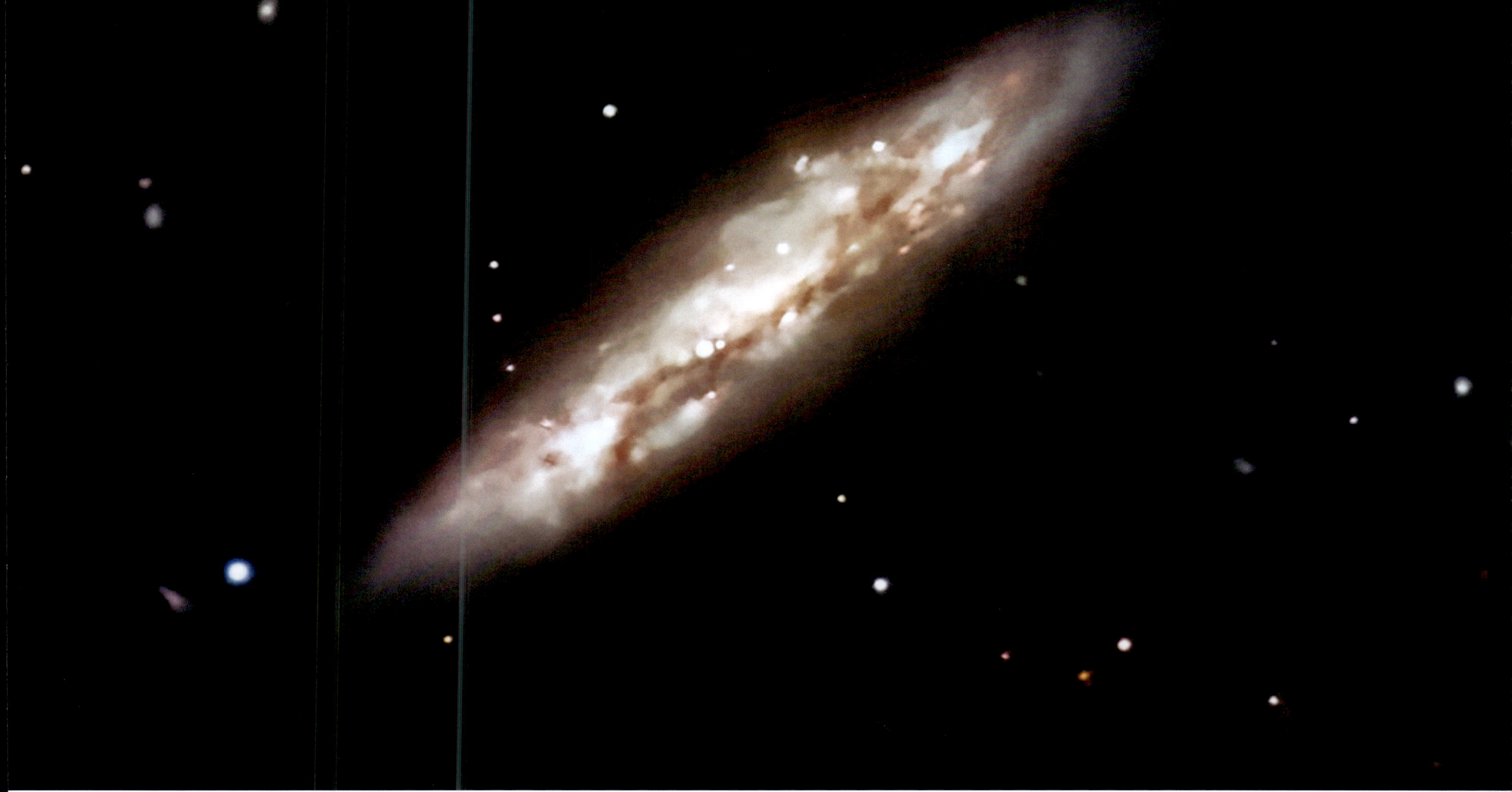

Messier 108 (The Surfboard Galaxy)
A spiral galaxy in the constellation Ursa Major
Approximate distance from Earth: 45 Million light-years

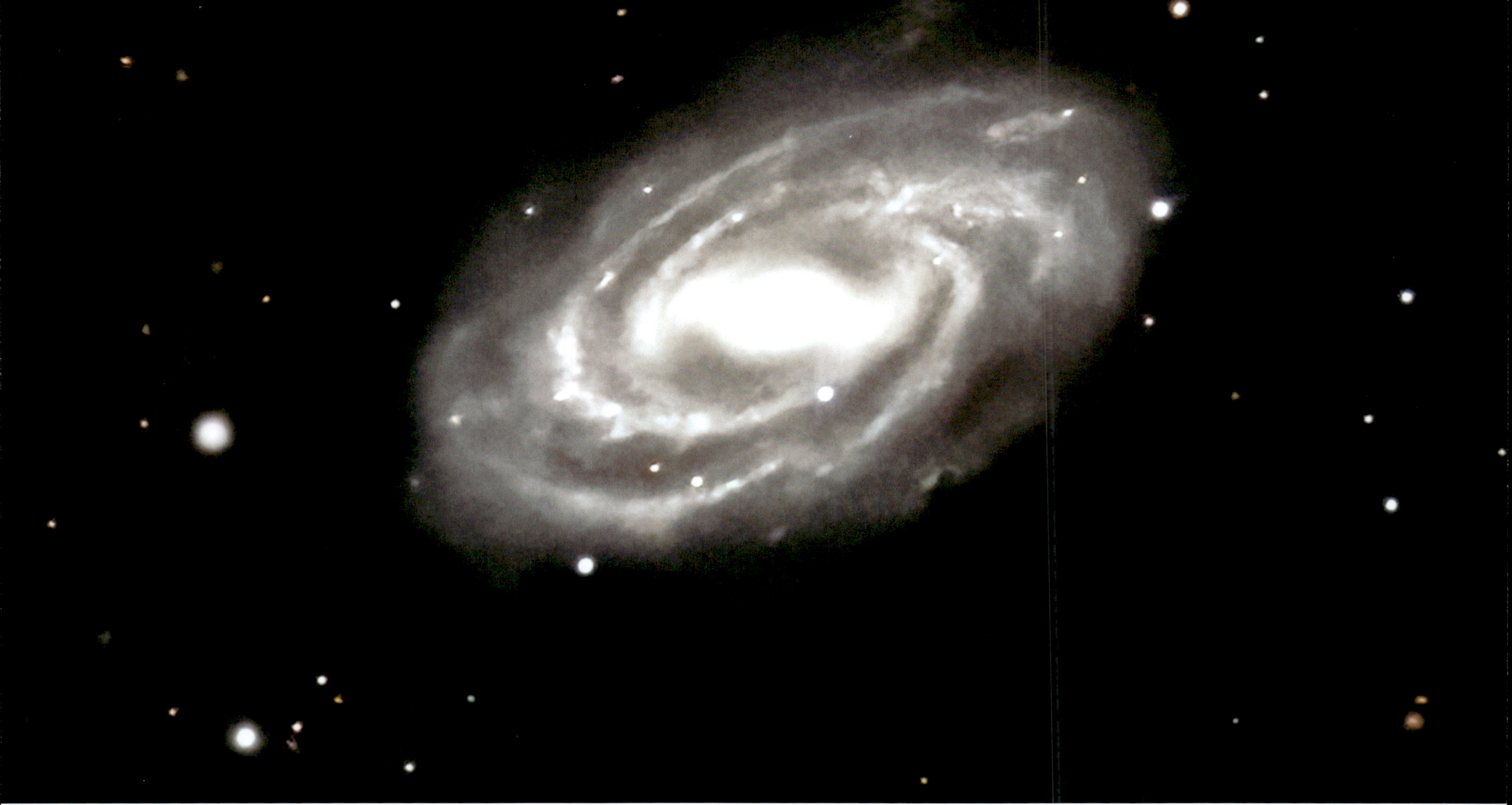

Messier 109 (The Theta Galaxy)
A barred spiral galaxy in the constellation Ursa Major
Approximate distance from Earth: 45 Million light-years

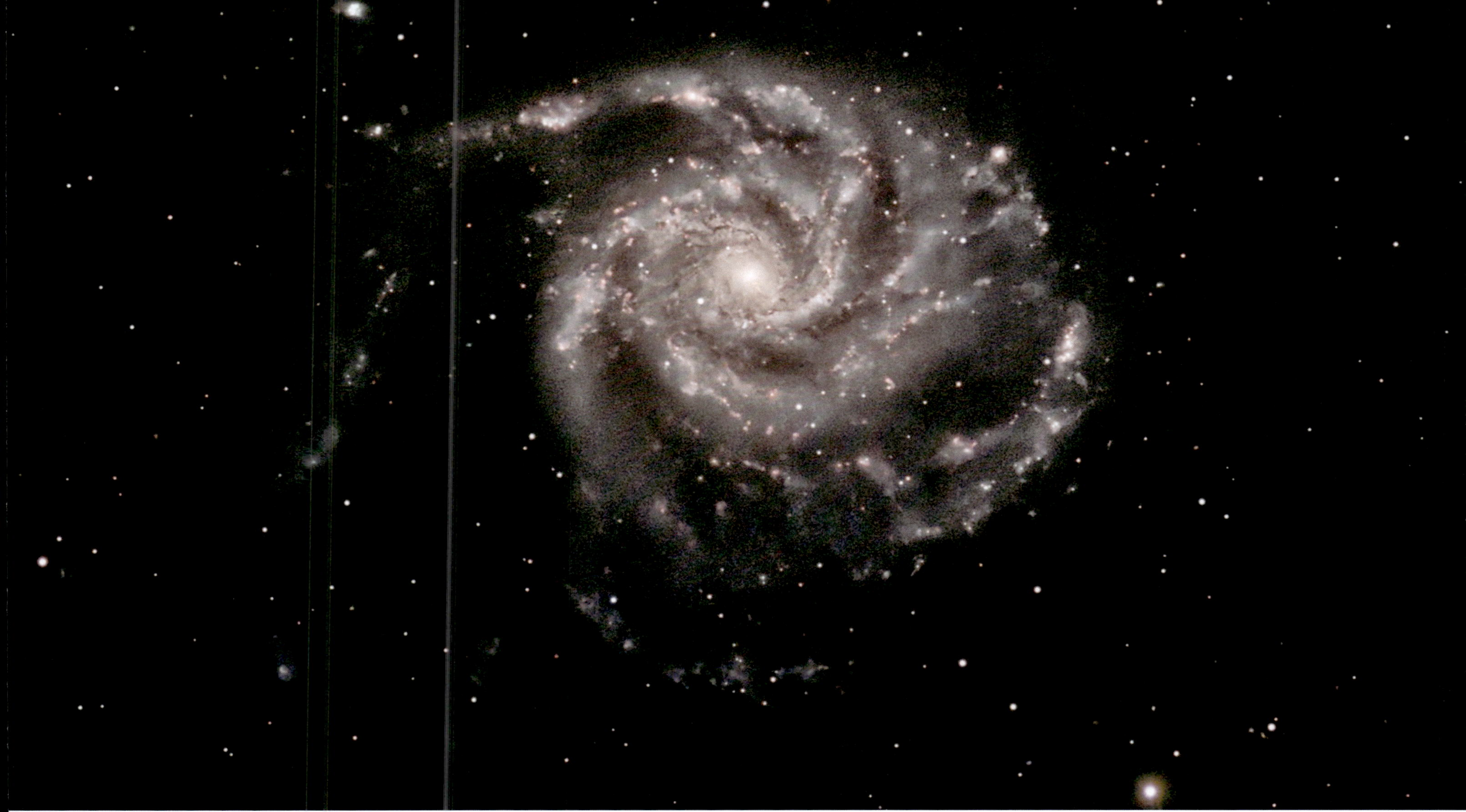

Messier 101 (The Pinwheel Galaxy)
A face-on spiral galaxy in the constellation Ursa Major
Approximate distance from Earth: 24 Million light-years

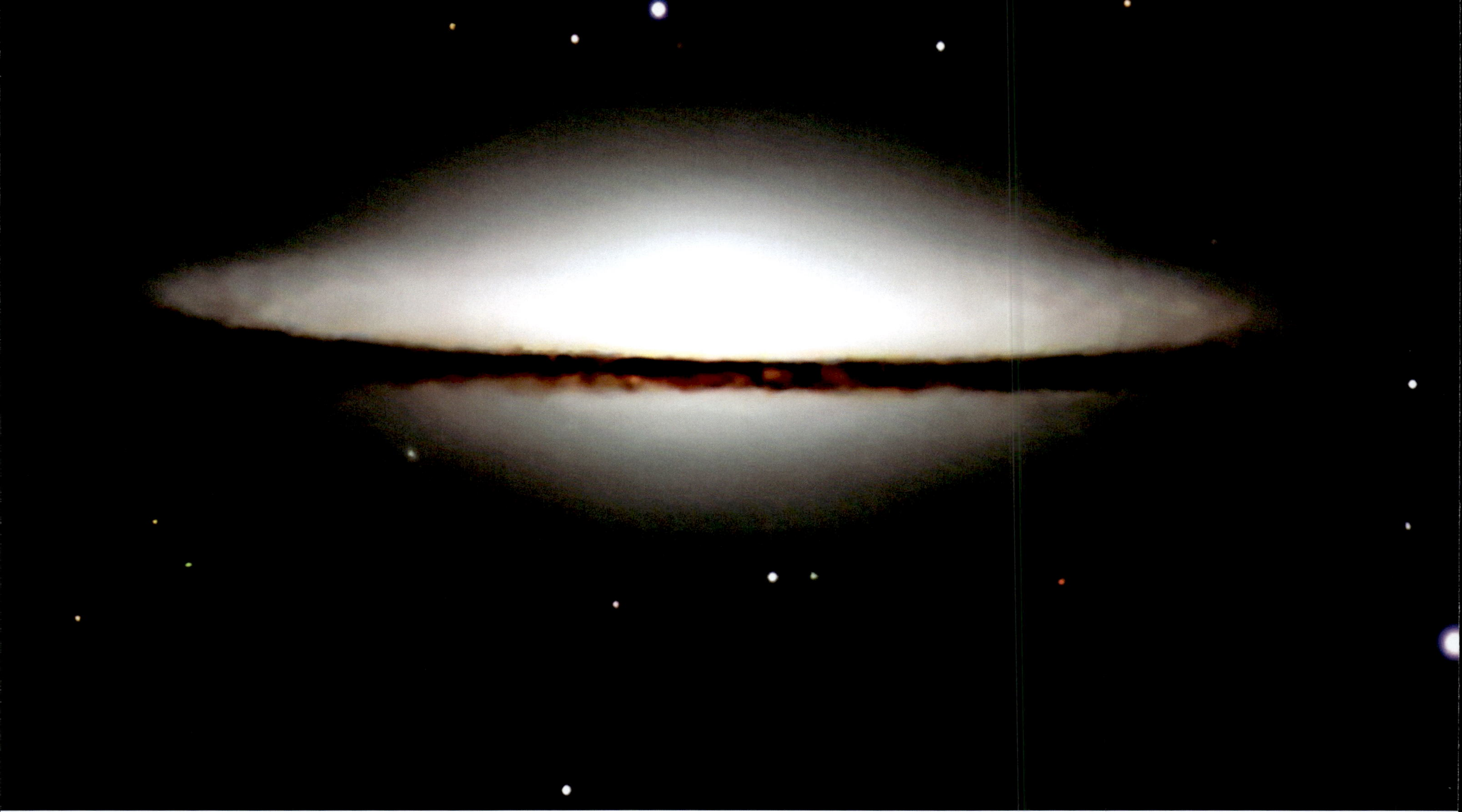

Messier 104 (The Sombrero Galaxy)
An edge-on lenticular galaxy in the constellation Virgo
Approximate distance from Earth: 31 Million light-years

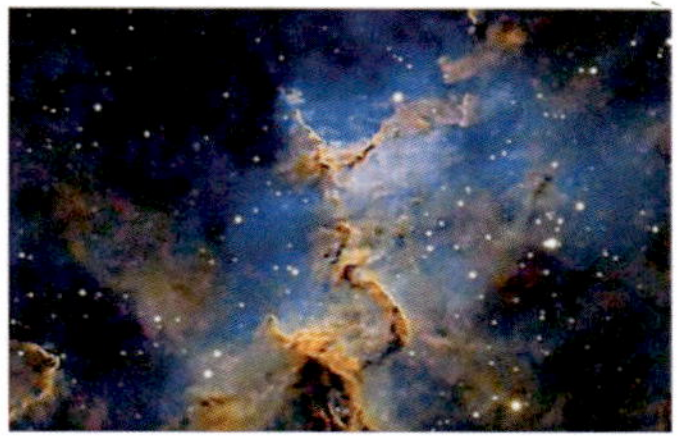

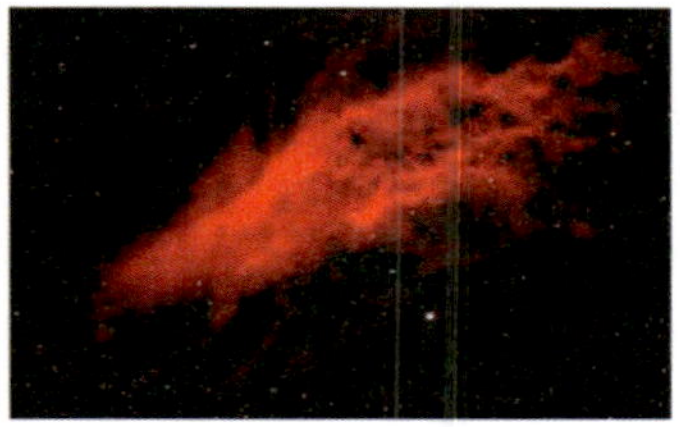

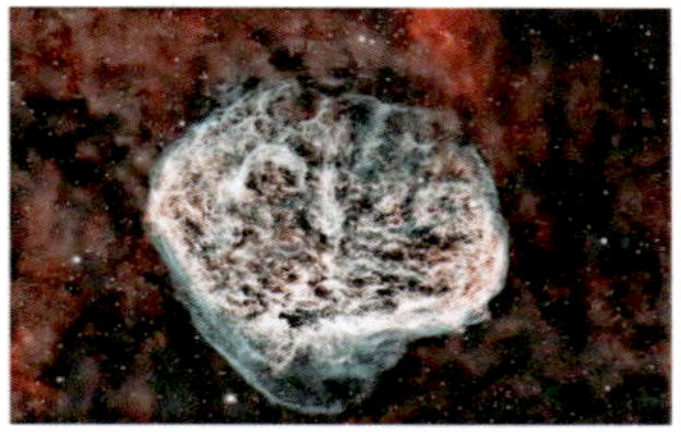

Nebulae Images

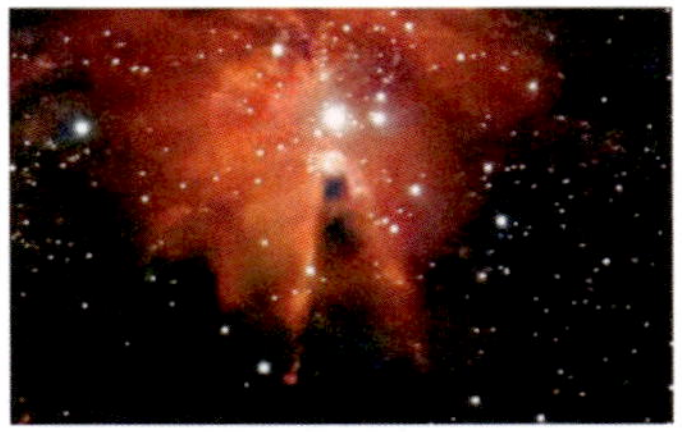

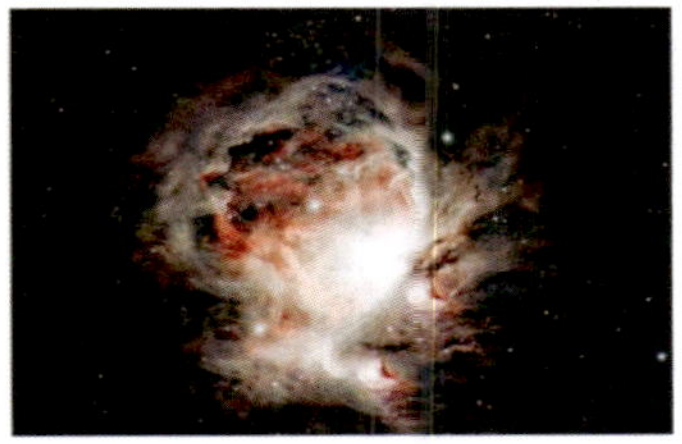

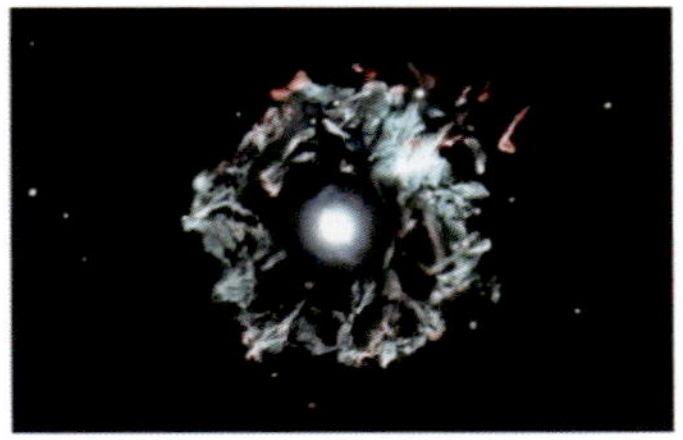

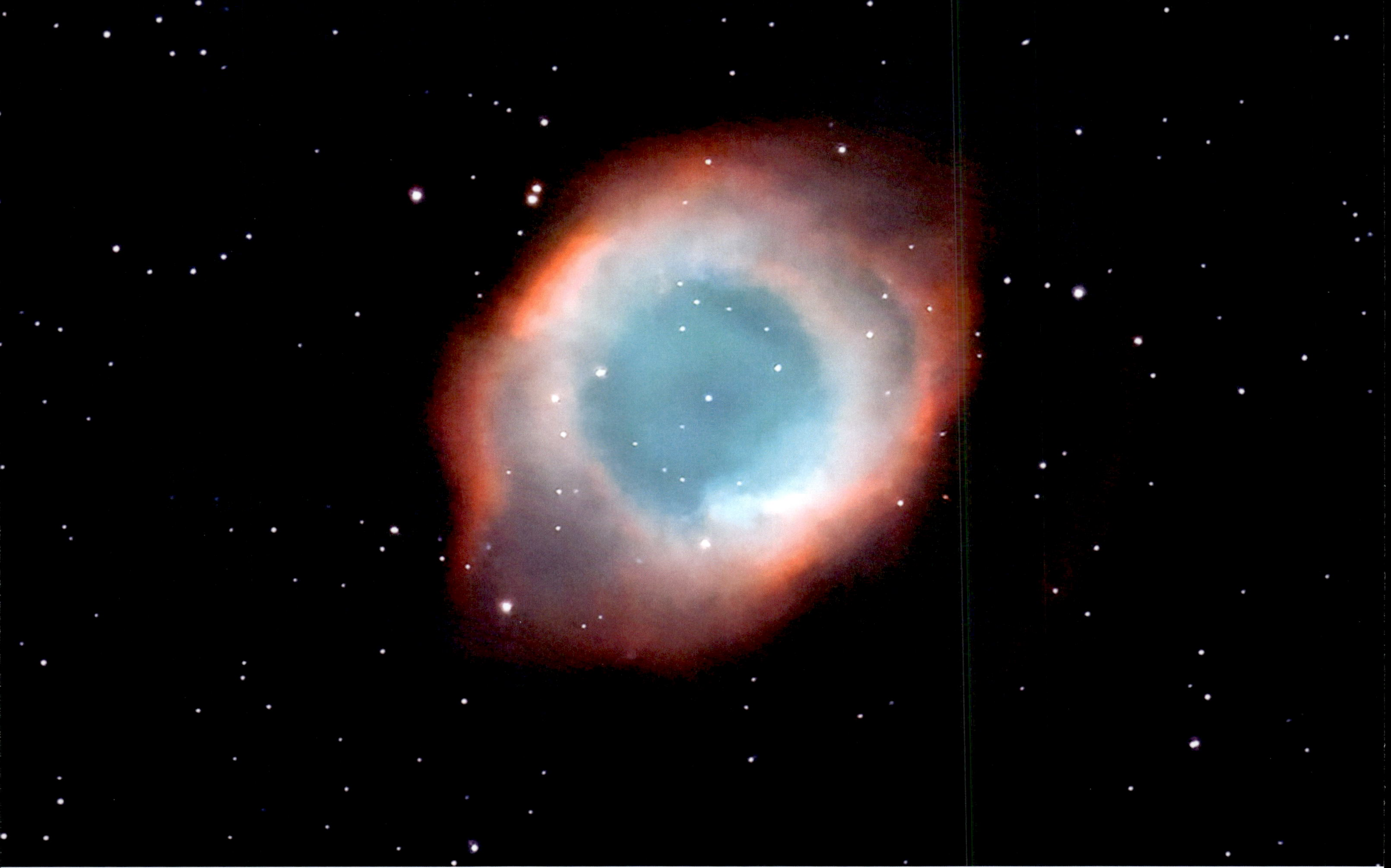

The Helix Nebula (NGC 7293)
A planetary nebula in the constellation Aquarius
Approximate distance from Earth: 650 light-years

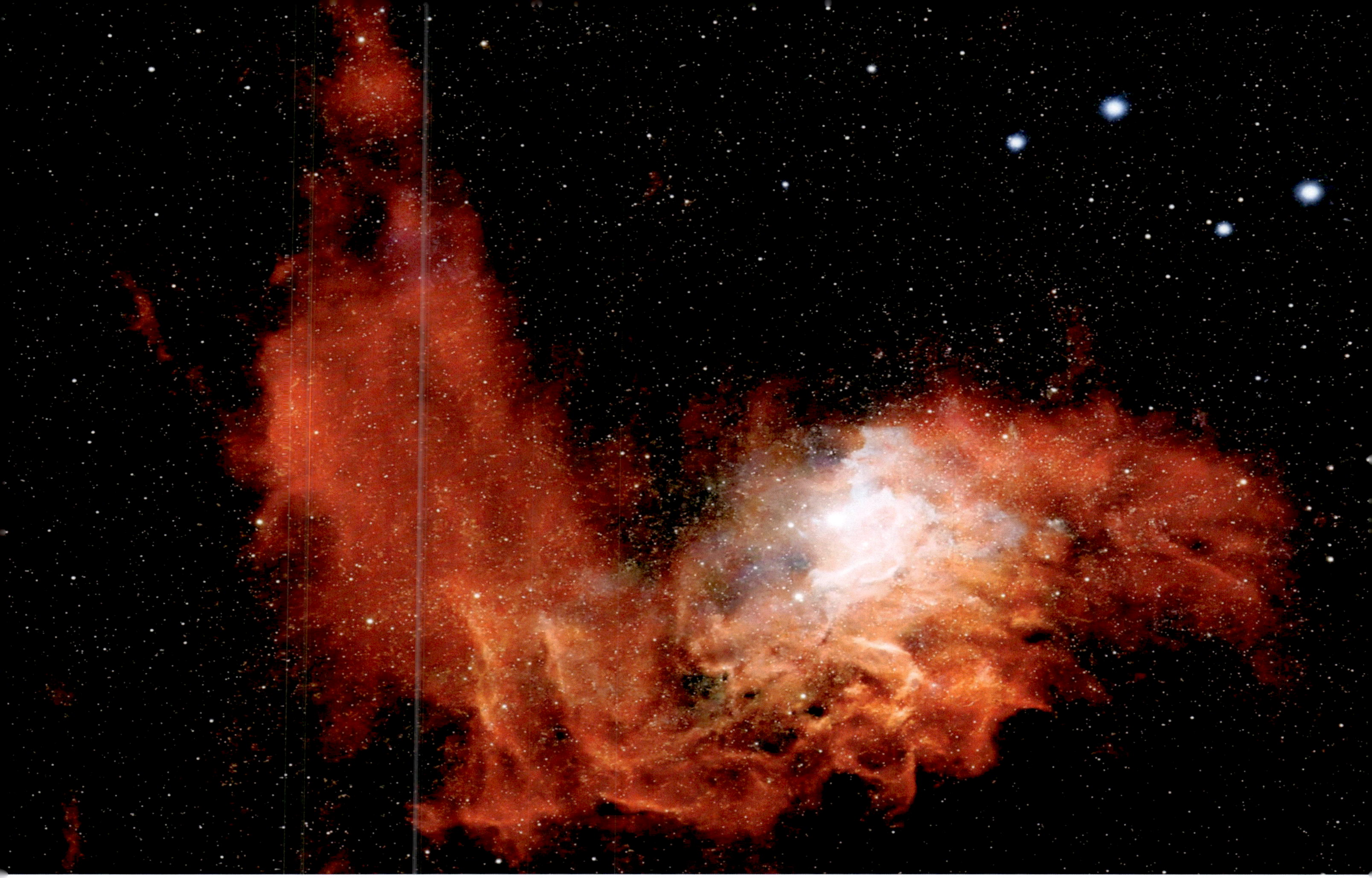

Flaming Star Nebula (IC 405)
An emission and reflection nebula in the constellation Auriga
Approximate distance from Earth: 1,500 light-years

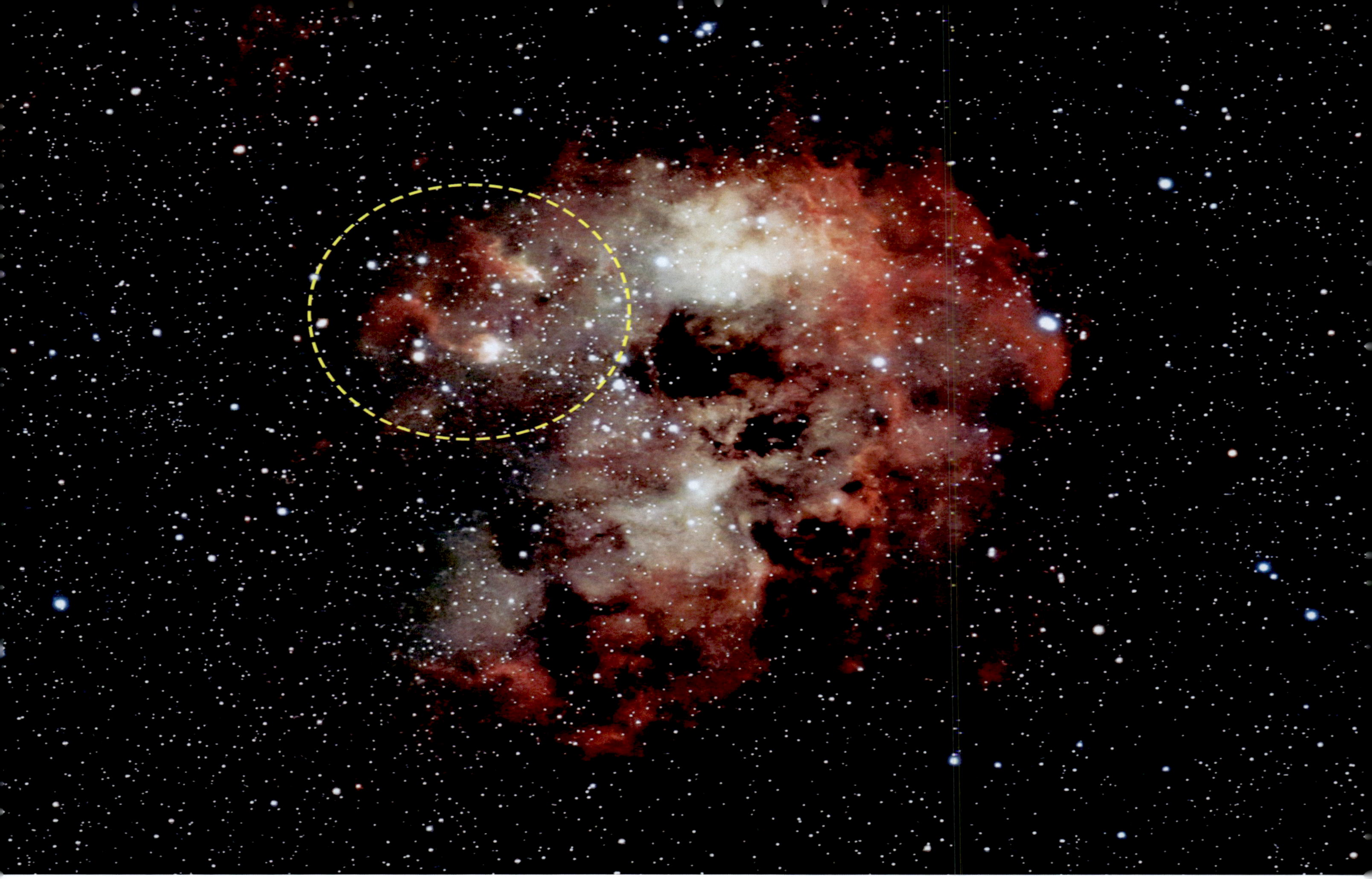

The Tadpole Nebula (IC 410)
An emission nebula in the constellation Auriga
Approximate distance from Earth: 12,000 light-years

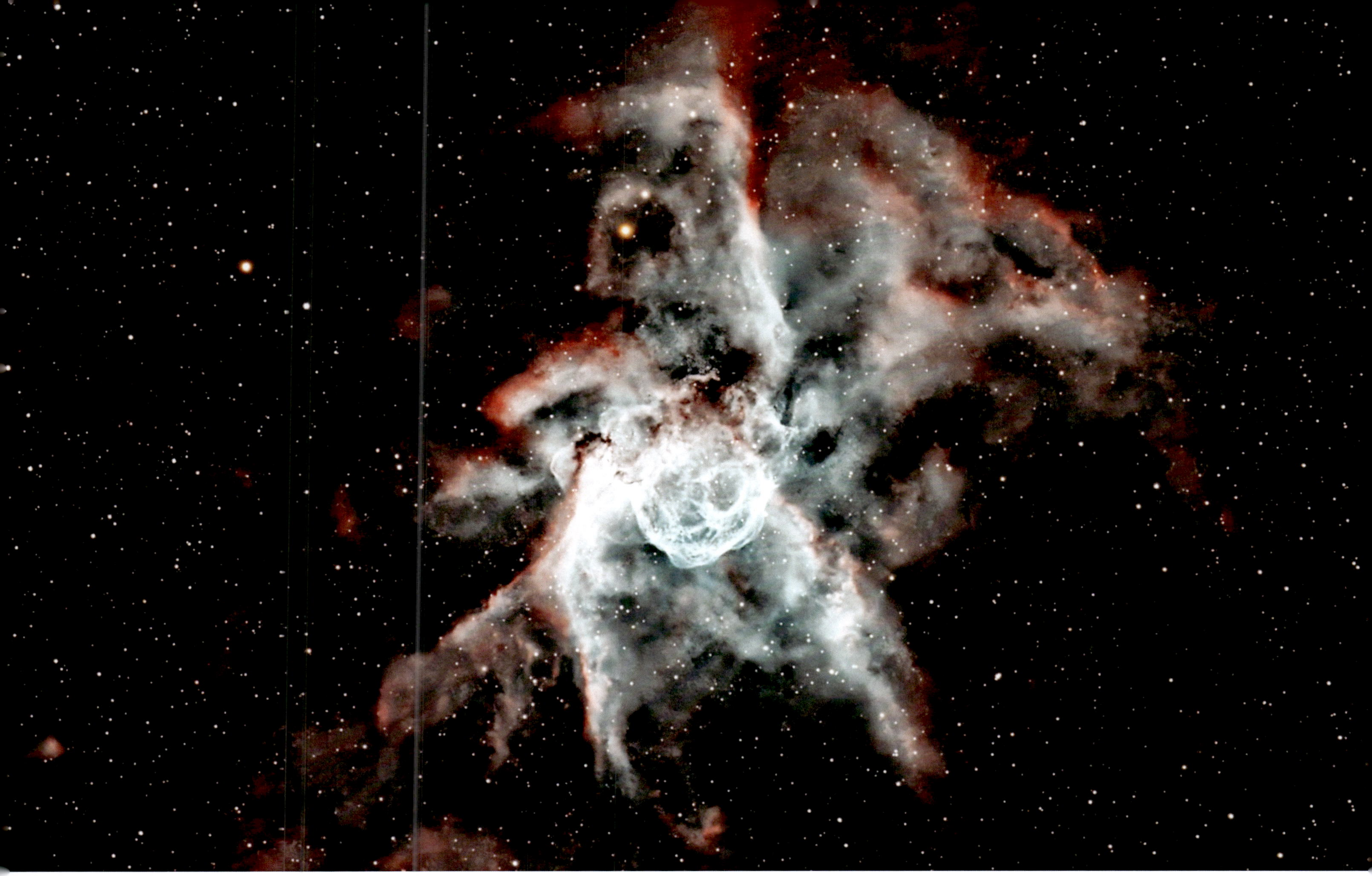

Thor's Helmet Nebula (NGC 2359)
An emission nebula in the constellation Canis Minor
Approximate distance from Earth: 11,700 light-years

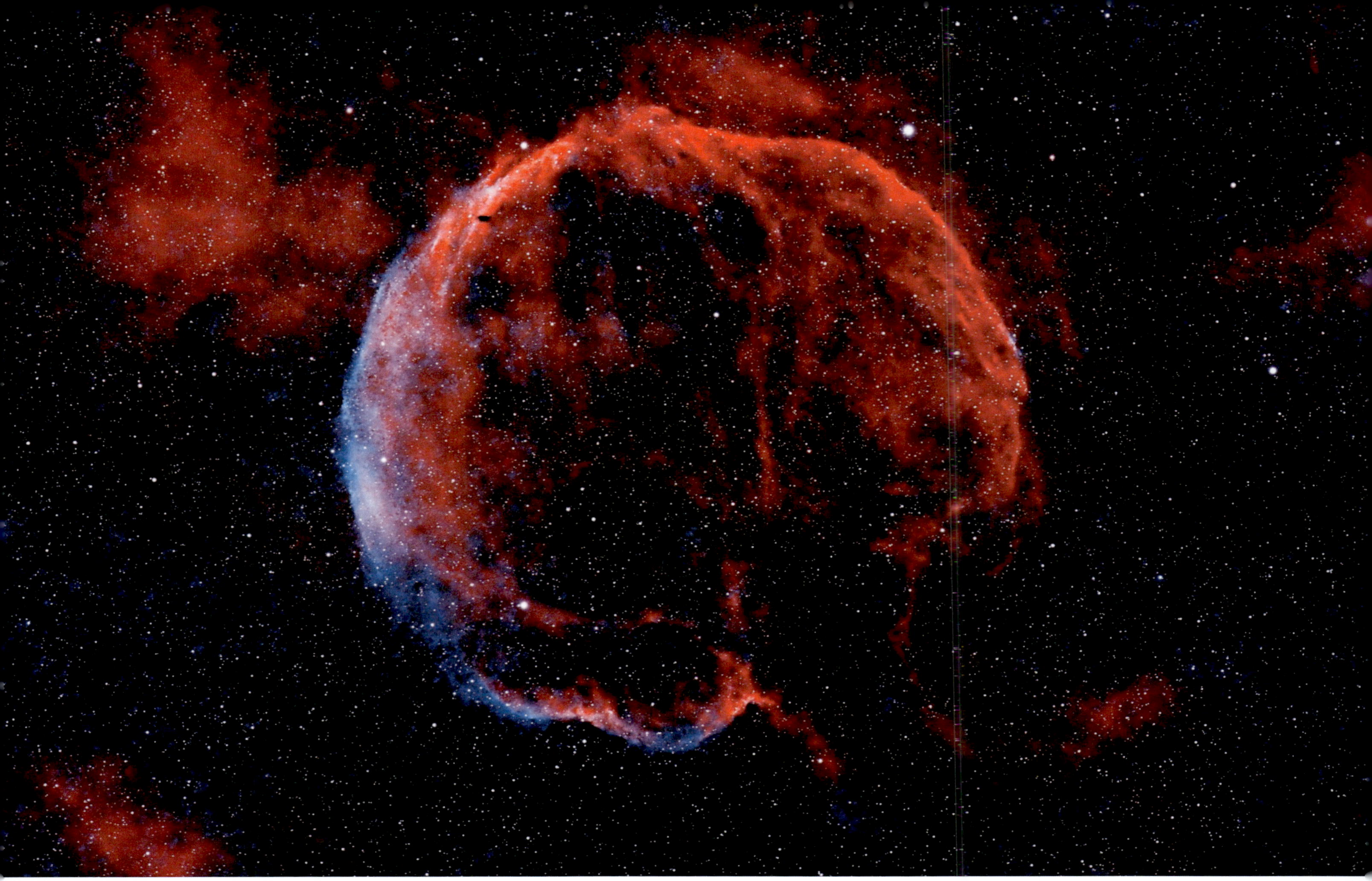

Popped Balloon Nebula (LBN 576)
A supernova remnant in the constellation Cassiopeia
Approximate distance from Earth: 8,000 light-years

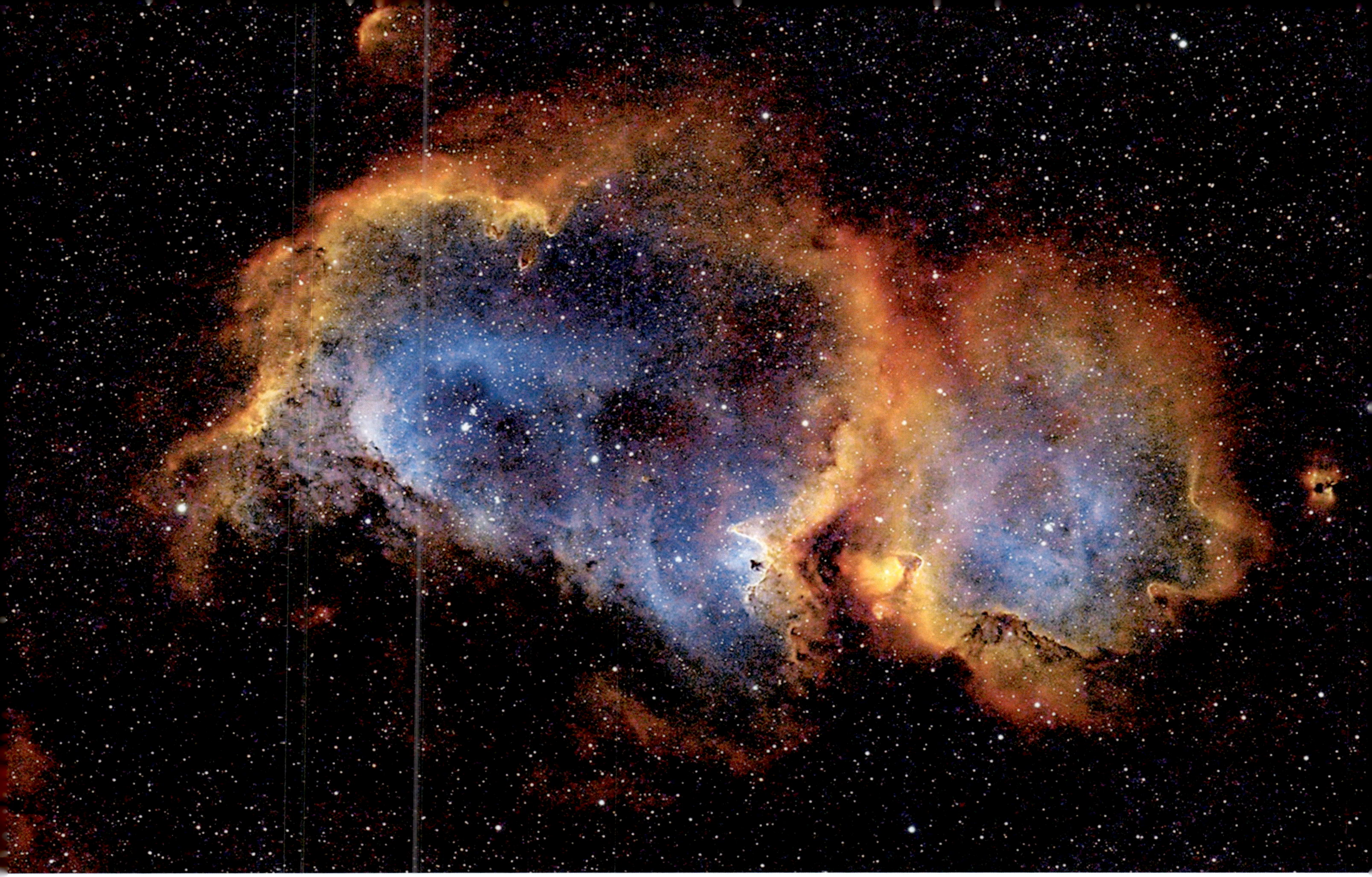

The Soul Nebula (IC 1848) – Emission Nebula
An emission nebula in the constellation Cassiopeia
Approximate distance from Earth: 7,500 light-years

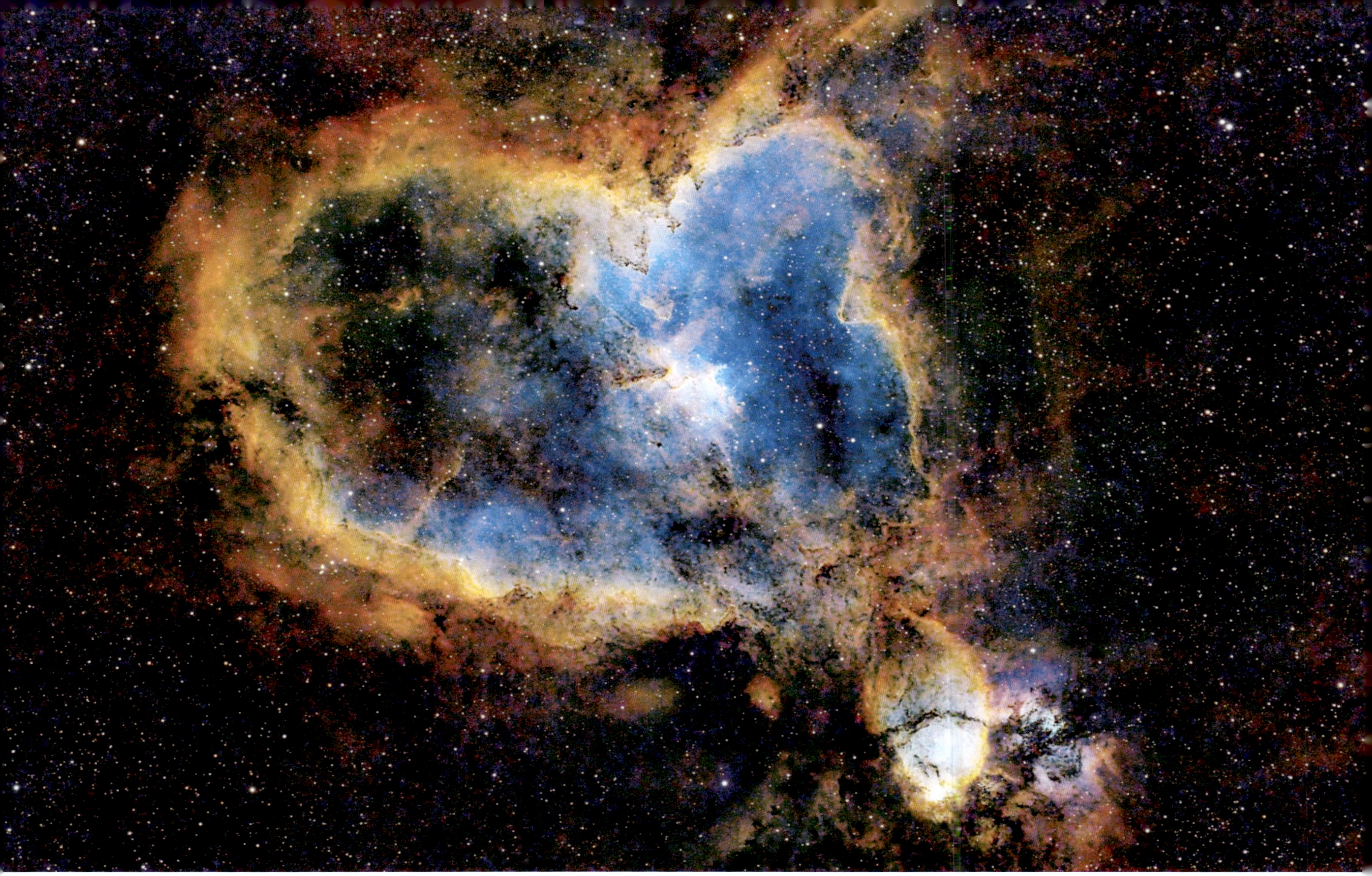

The Heart Nebula (IC 1805)
An emission nebula in the constellation Cassiopeia
Approximate distance from Earth: 7,500 light-years

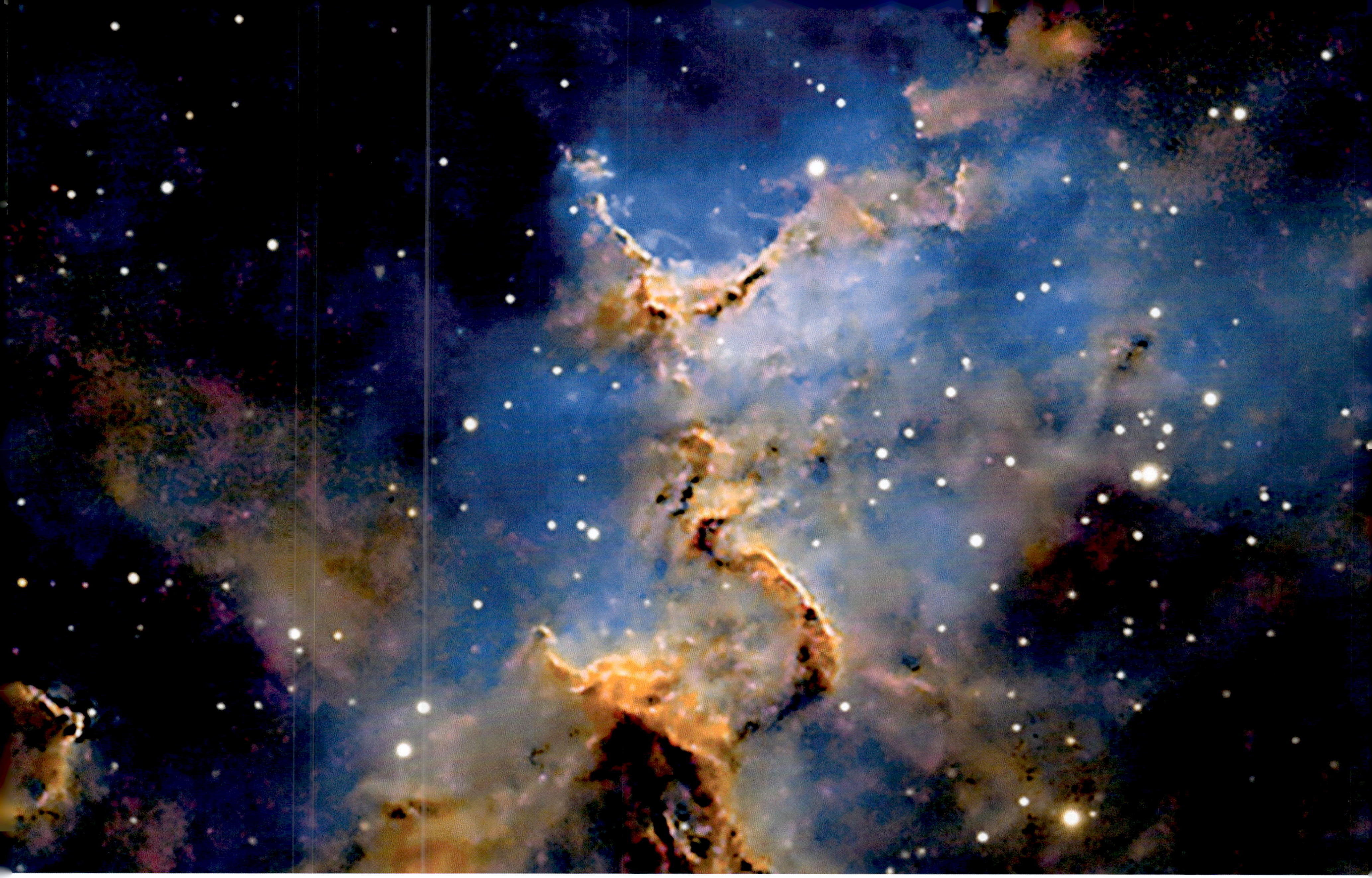

Melotte 15 — The Center of the Heart Nebula
A young open star cluster in an active stellar nursery / star-forming region in the constellation Cassiopeia
Approximate distance from Earth: 7,500 light-years

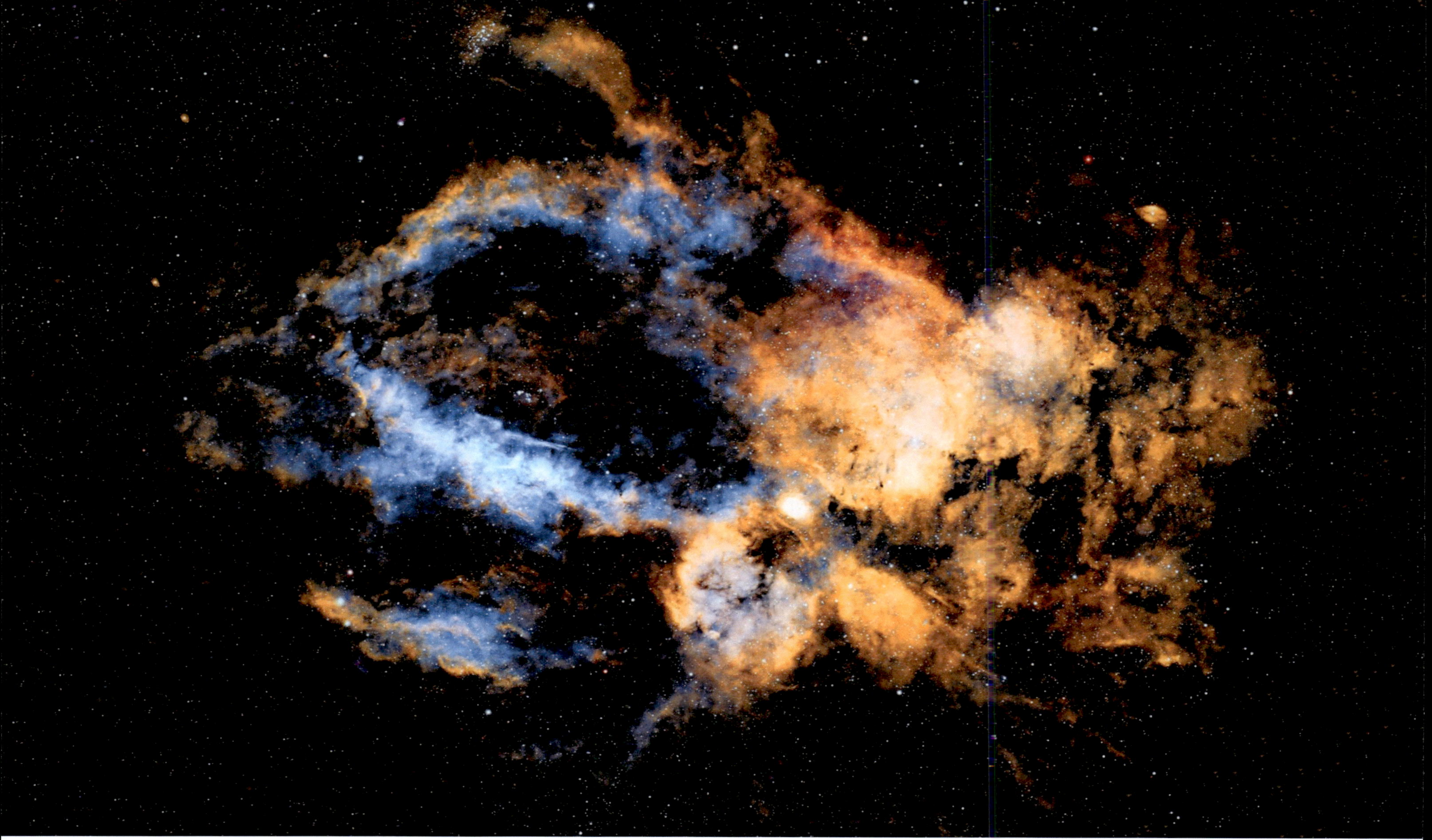

Lobster Claw Nebula — 2 panel mosaic (Sh2-157)
An emission nebula in the constellation Cassiopeia
Approximate distance from Earth: 11,000 light-years

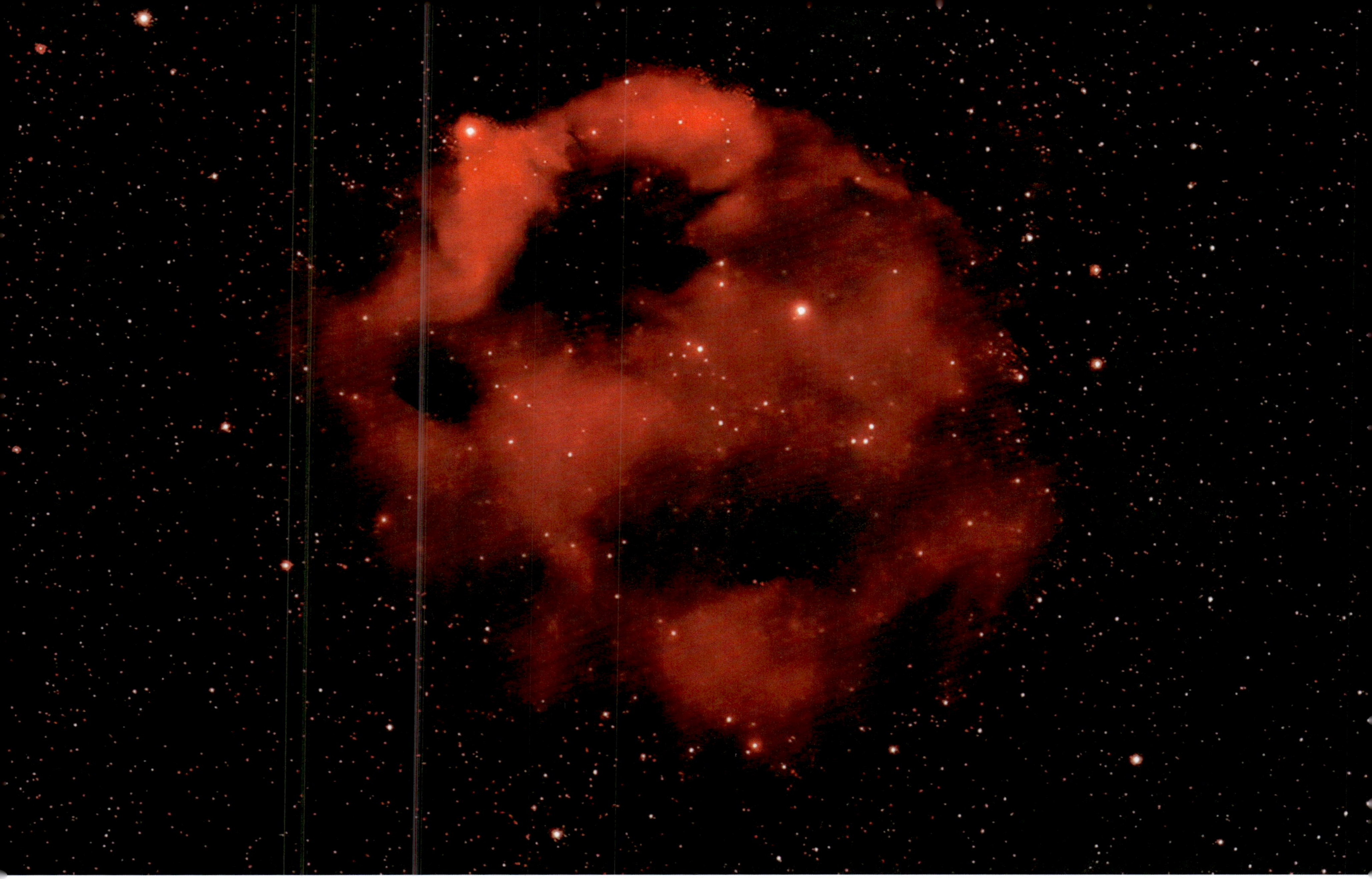

The Phantom of the Opera Nebula (Sharpless 2-173)
An emission nebula in the constellation Cassiopeia
Approximate distance from Earth: 10,000 light-years

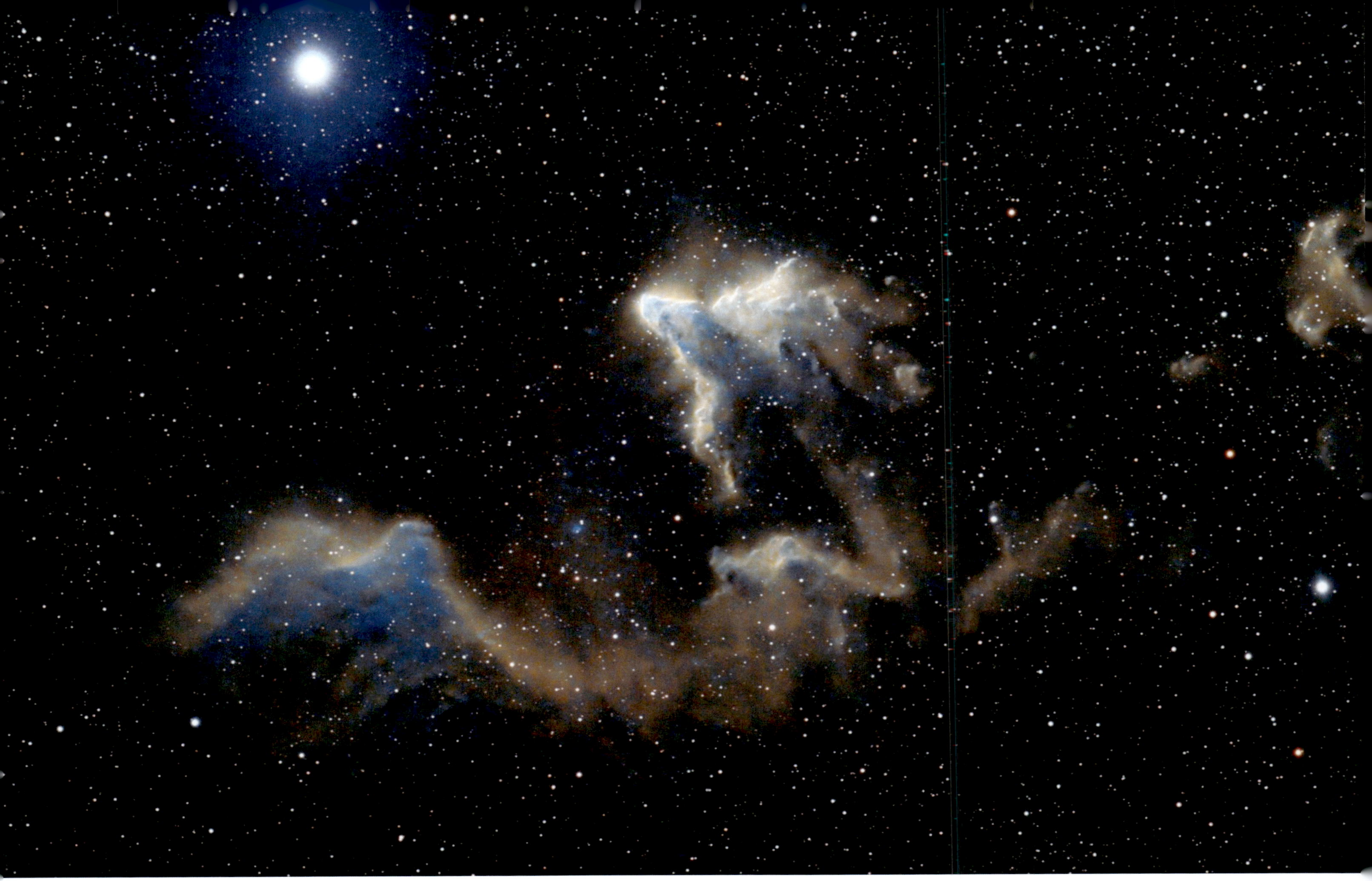

The Ghost of Cassiopeia Nebula (IC 63)
An emission and refection nebula in the constellation Cassiopeia
Approximate distance from Earth: 550 light-years

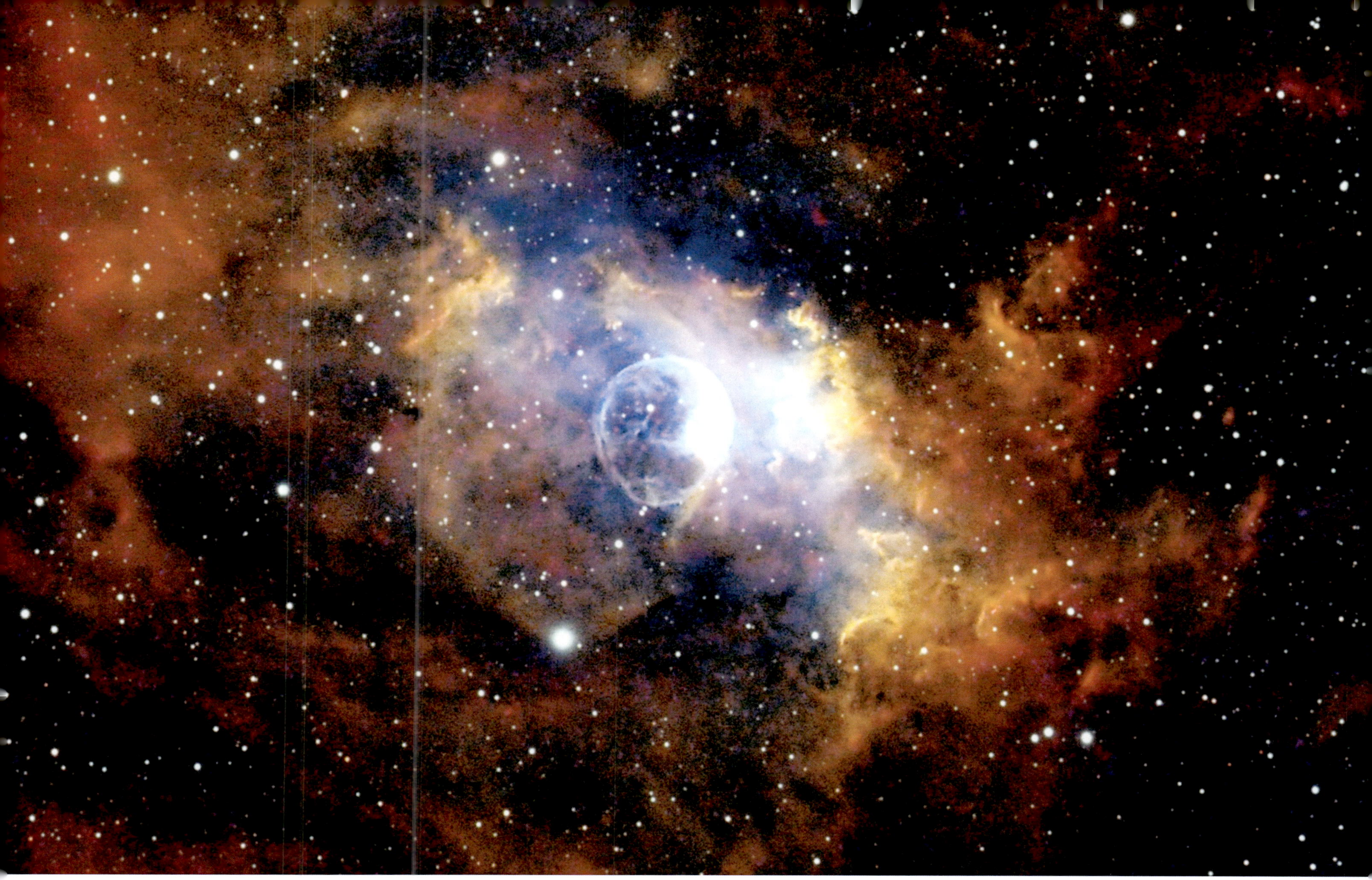

The Bubble Nebula (NGC 7635) – Emission Nebula
An emission nebula in the constellation Cassiopeia
Approximate distance from Earth: 7,100 light-years

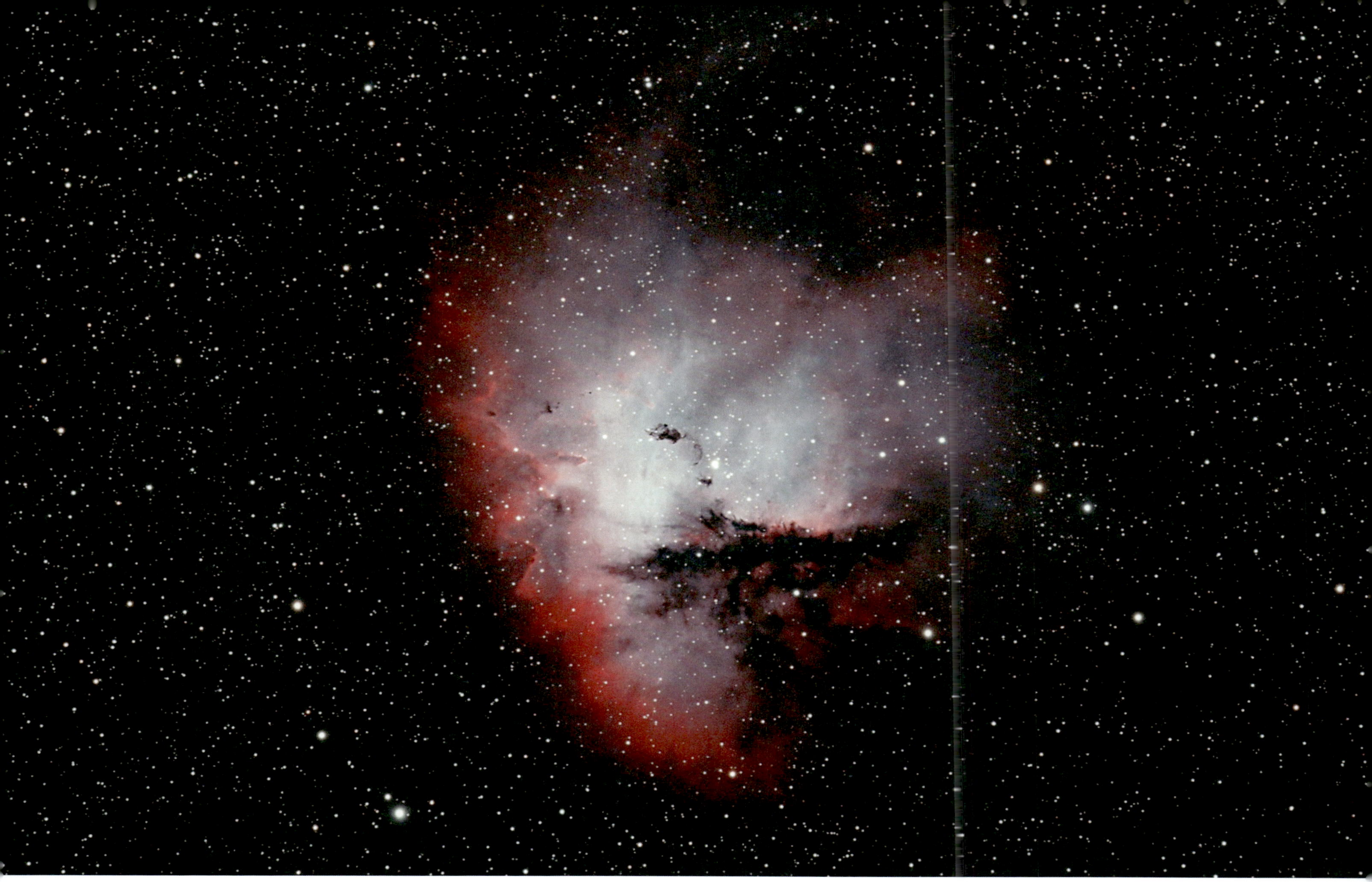

Pacman Nebula (NGC 281)
An emission nebula in the constellation Cassiopeia
Approximate distance from Earth: 9,500 light-years

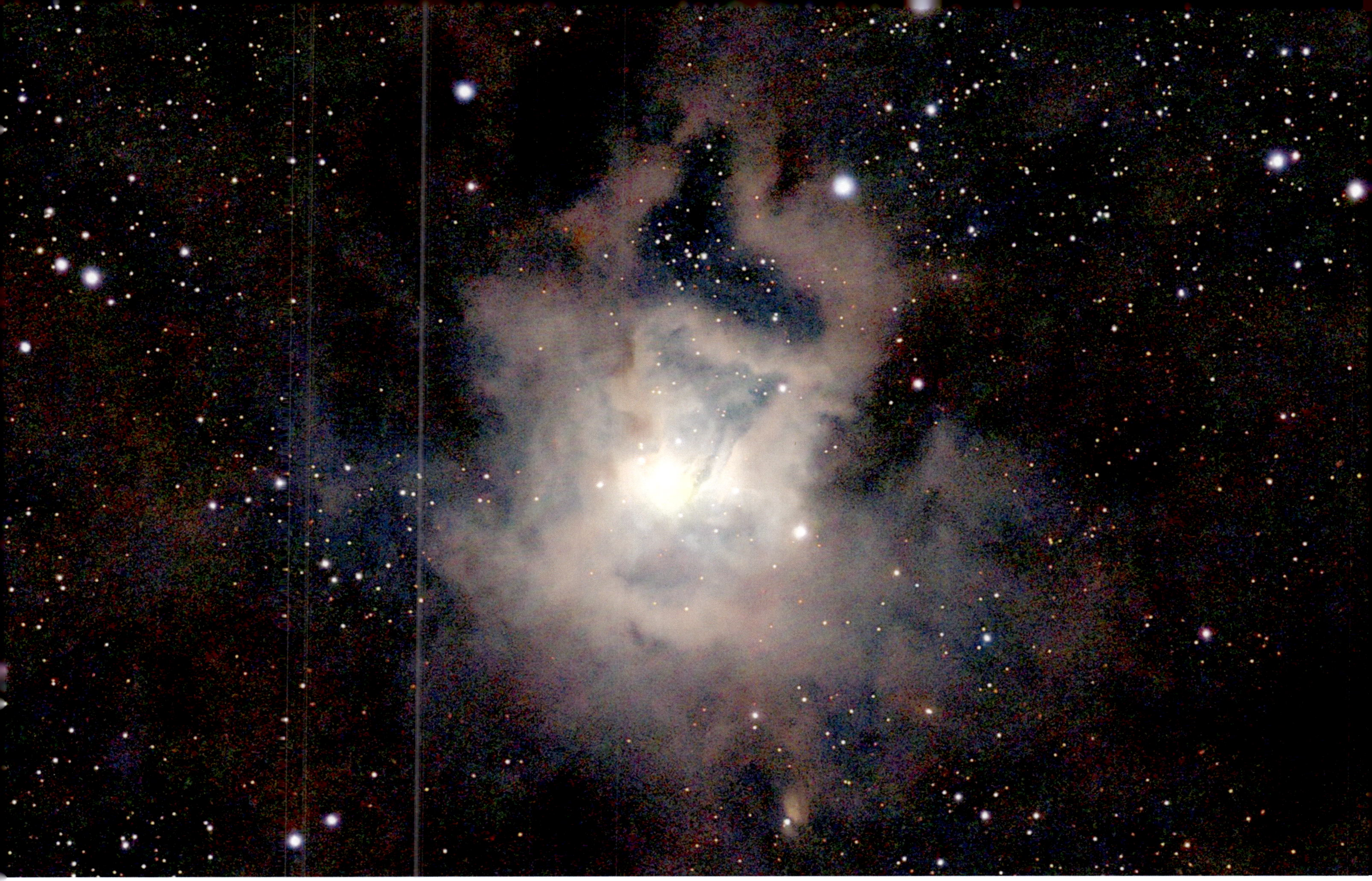

The Iris Nebula (NGC 7023)
A reflection nebula in the constellation Cepheus
Approximate distance from Earth: 1,300 light-years

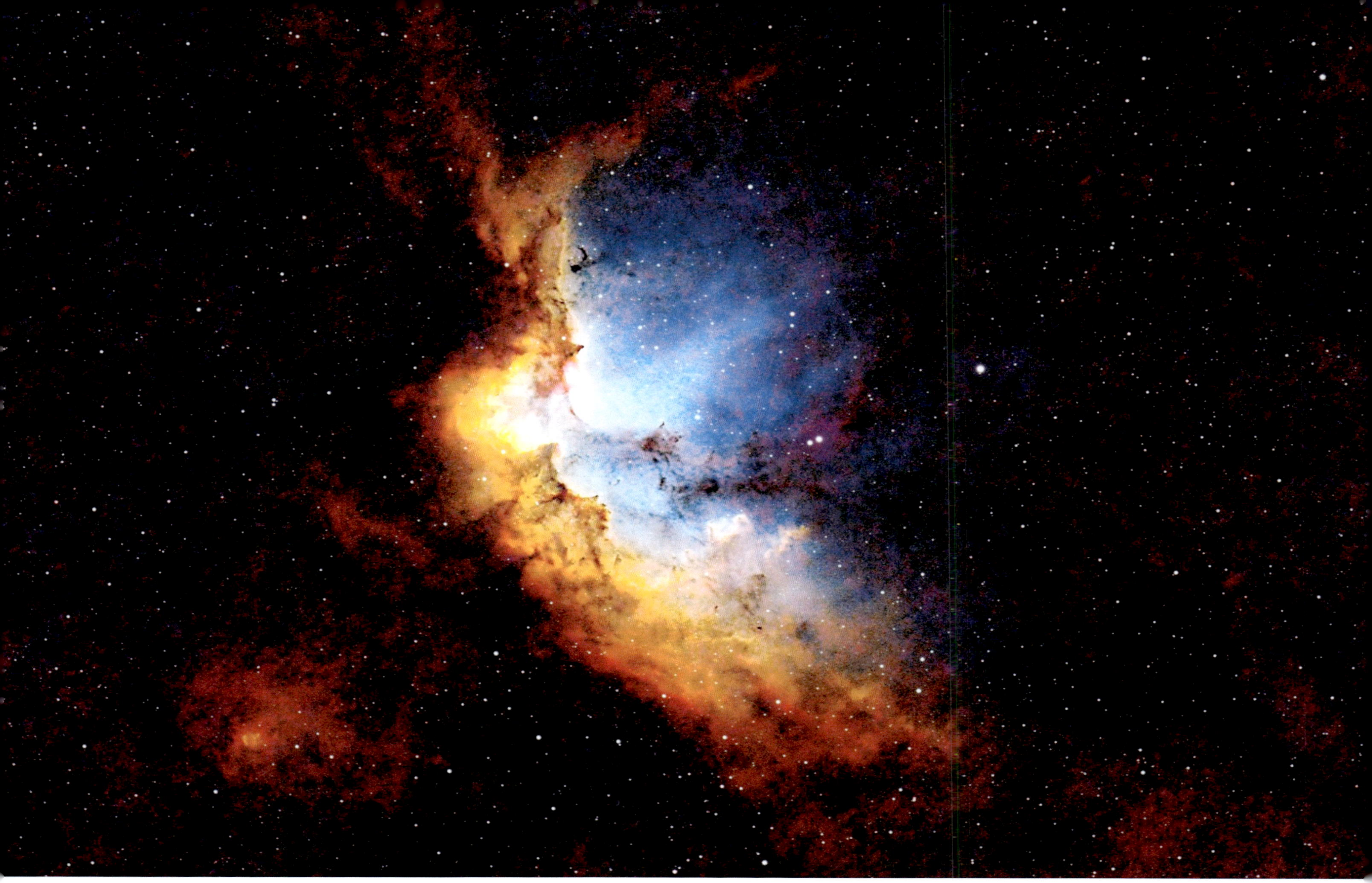

The Wizard Nebula (Sh2-142)
An emission nebula / star-forming region in the constellation Cepheus
Approximate distance from Earth: 7,200 light-years

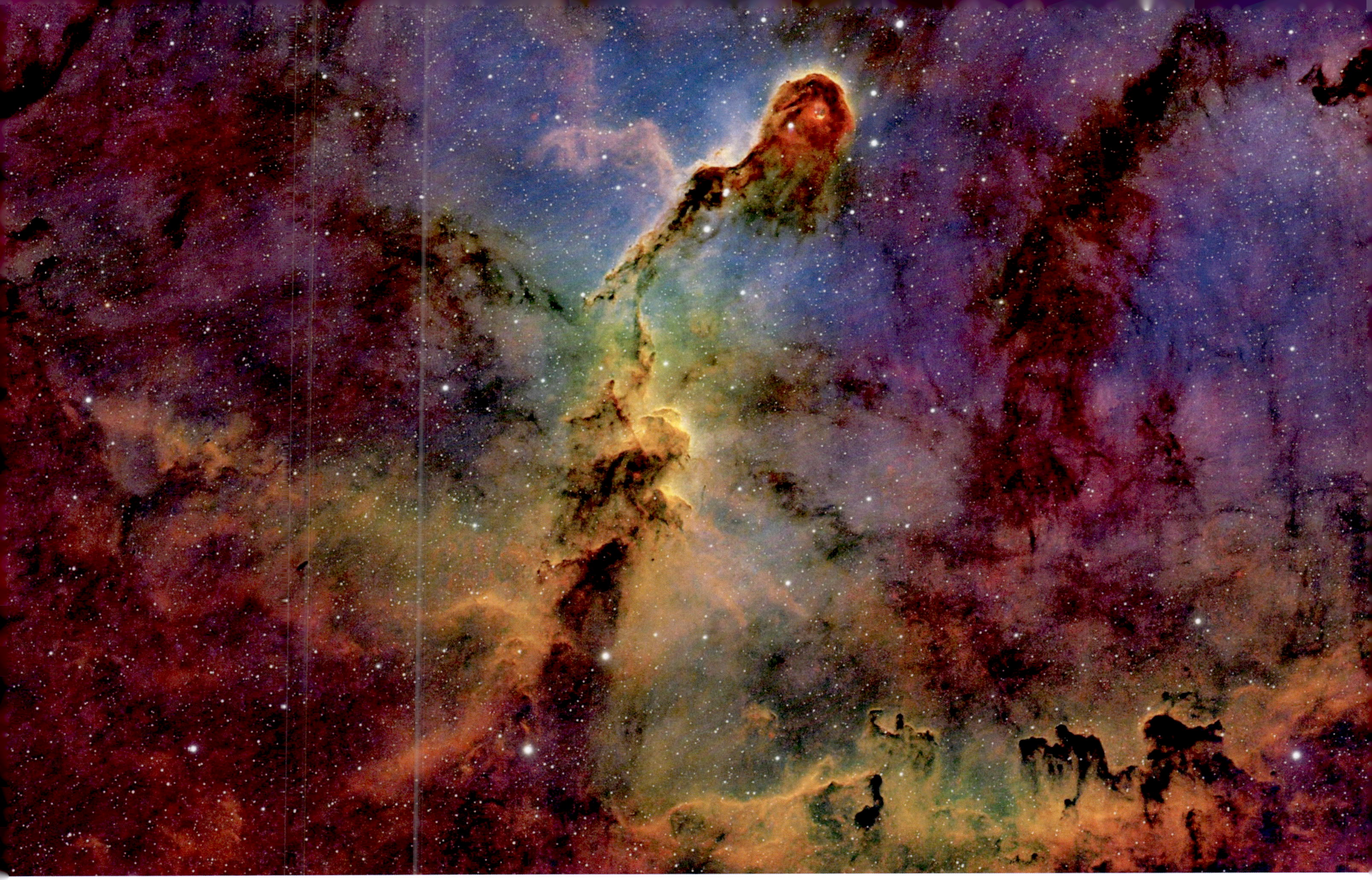

The Elephant Trunk Nebula (IC 1396)
An emission nebula / star-forming region in the constellation Cepheus
Approximate distance from Earth: 2,400 light-years

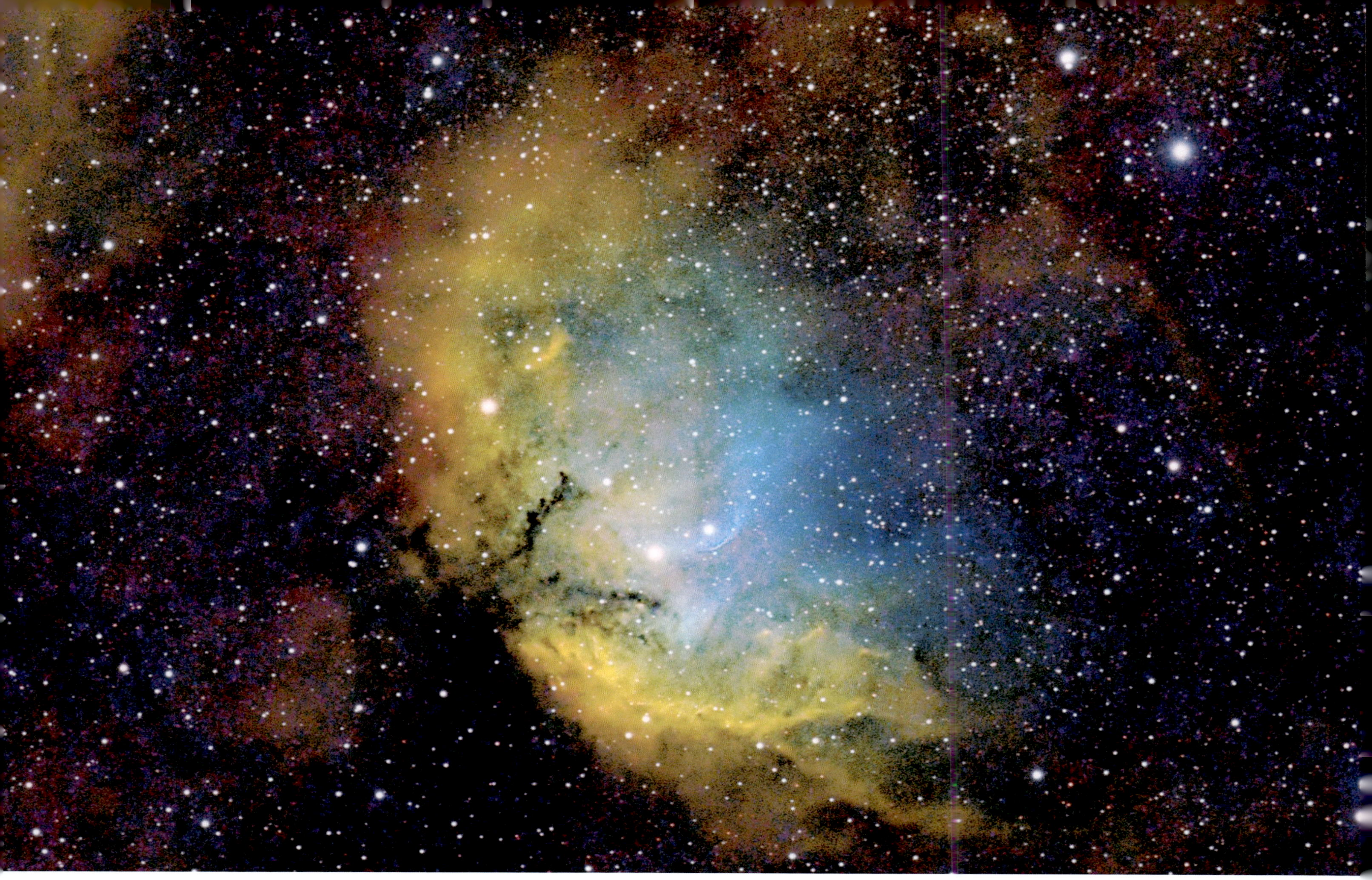

Tulip Nebula (Sh2-101)
An emission nebula in the constellation Cygnus
Approximate distance from Earth: 6,000 light-years

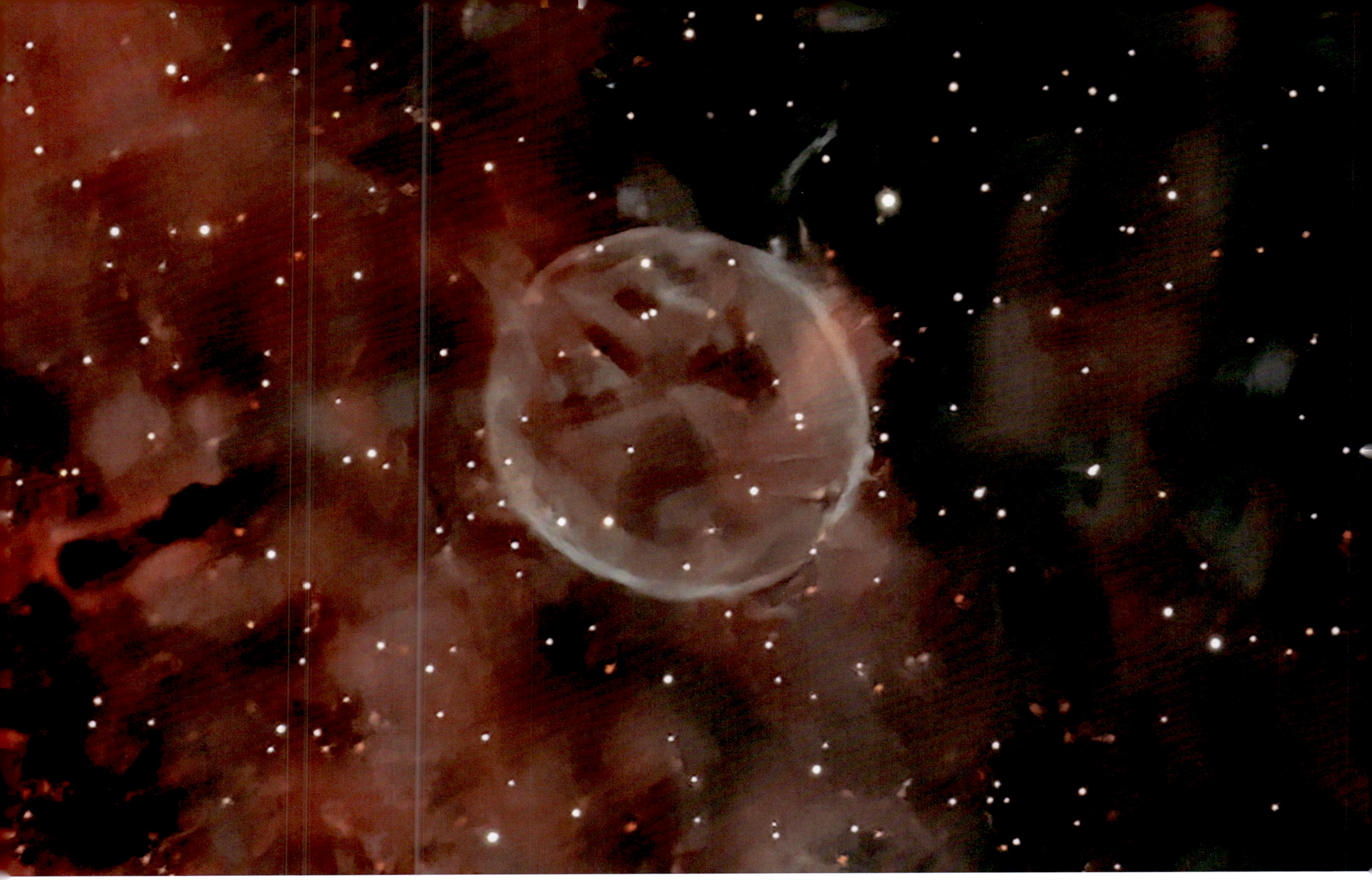

Soap Bubble Nebula
A planetary nebula in the constellation Cygnus
Approximate distance from Earth: 11,000 light-years

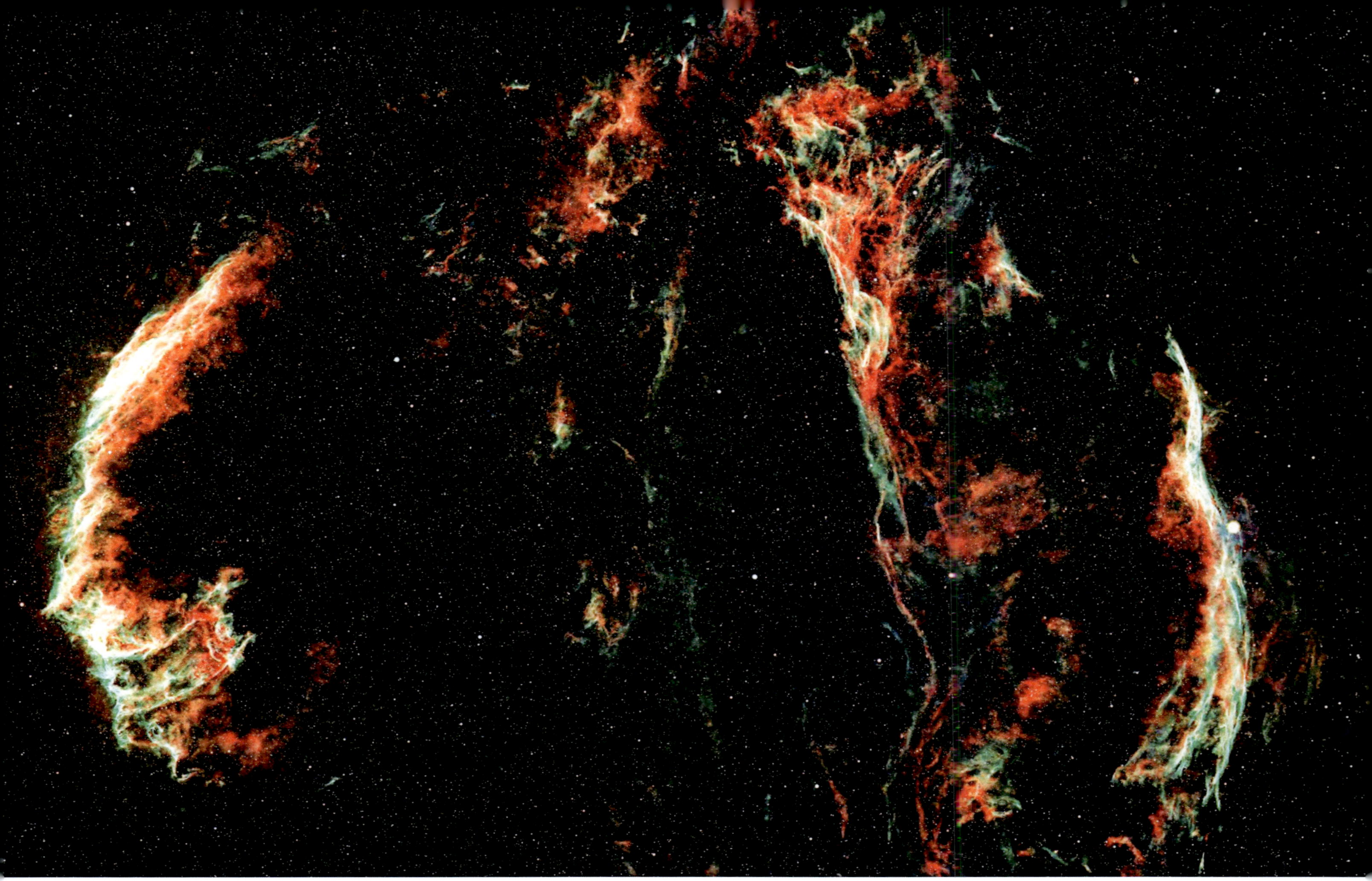

The Veil Nebula Full (Cygnus Loop)
A supernova remnant in the constellation Cygnus
Approximate distance from Earth: 2,400 light-years

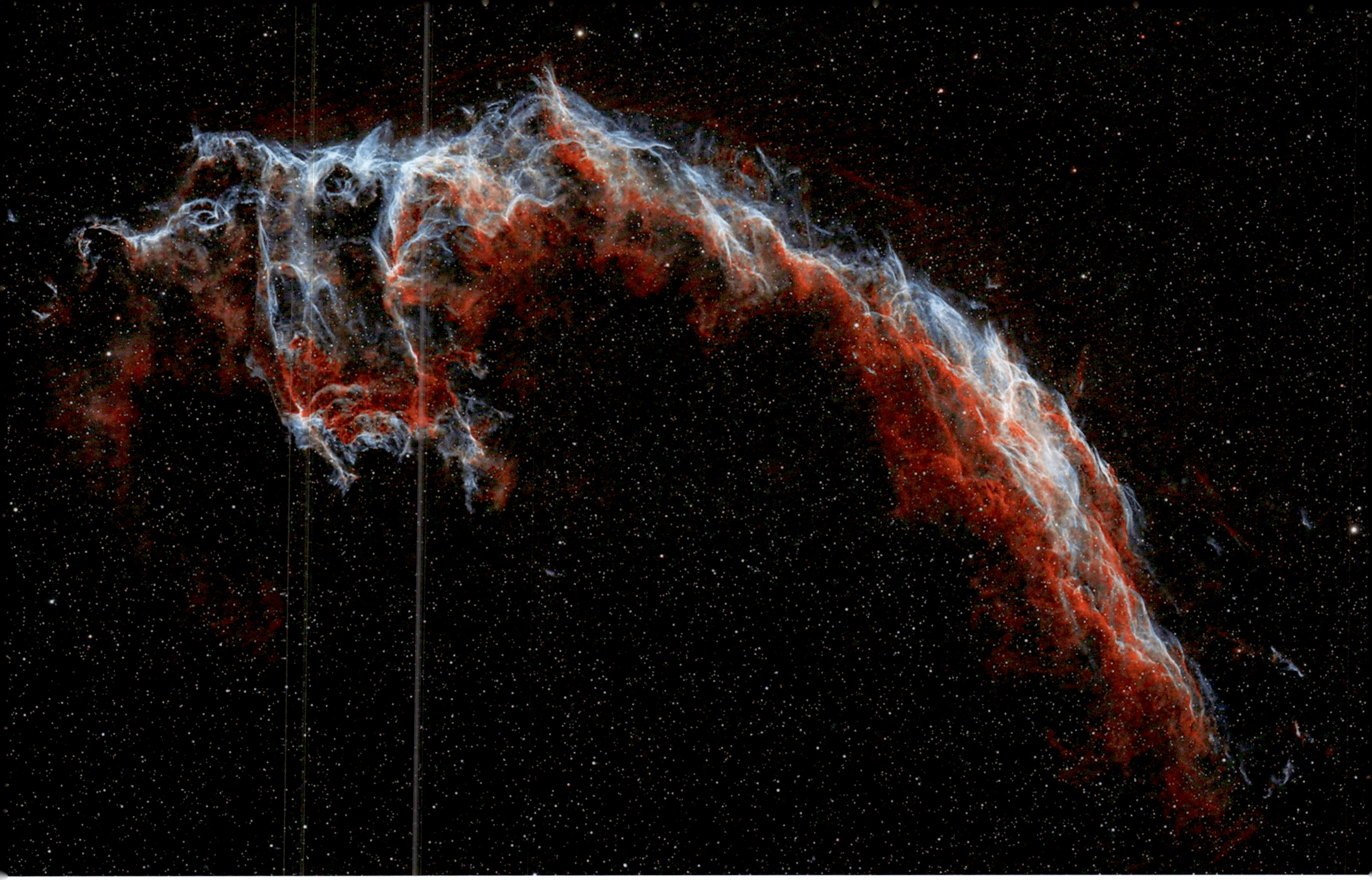

The Veil Nebula East (NGC 6992 / NGC 6995)
A bright part of the Cygnus Loop supernova remnant in the constellation Cygnus
Approximate distance from Earth: 2,400 light-years

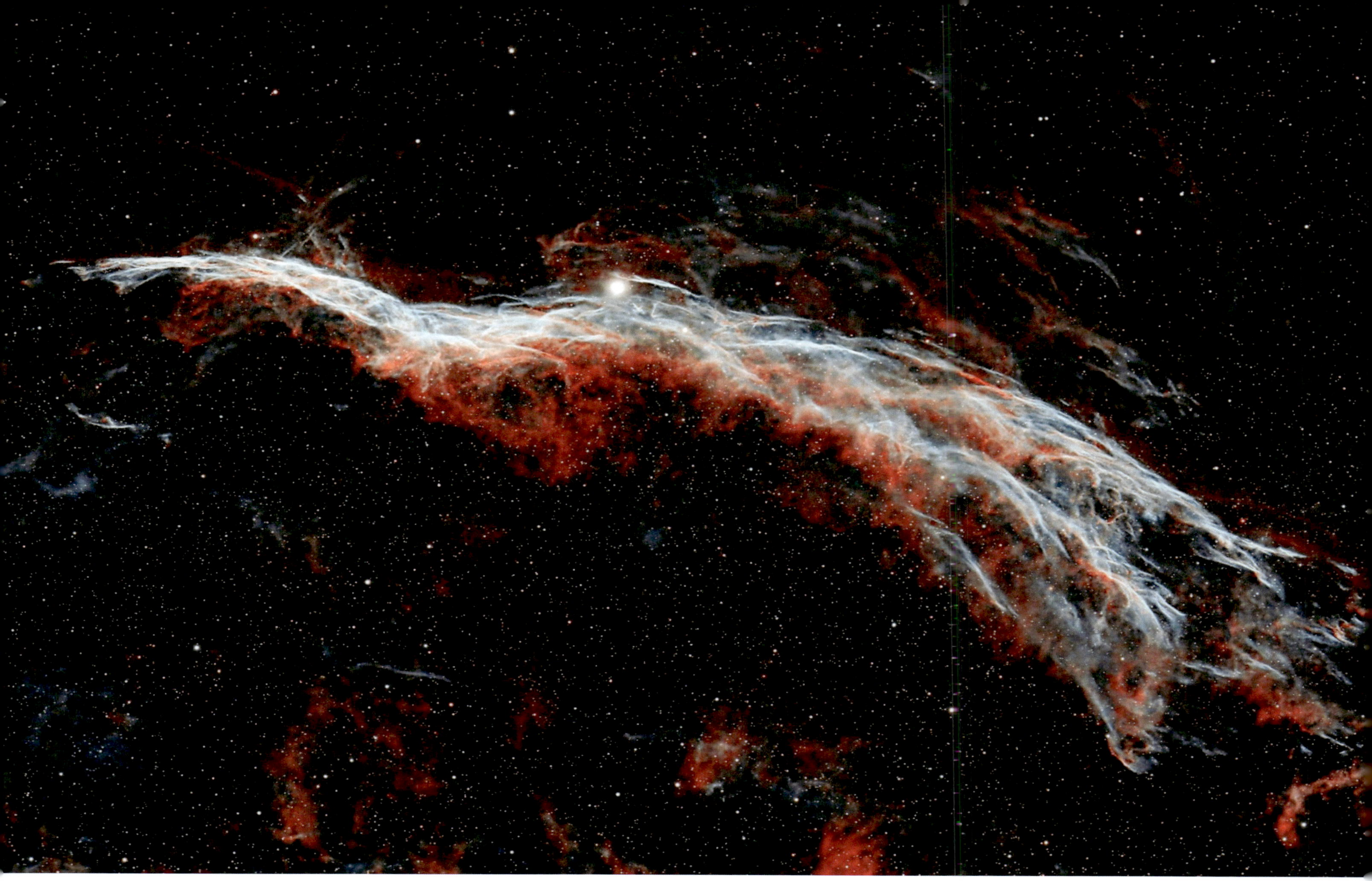

The Veil Nebula West (NGC 6960 – The Witch's Broom)
A bright part of the Cygnus Loop supernova remnant in the constellation Cygnus
Approximate distance from Earth: 2,400 light-years

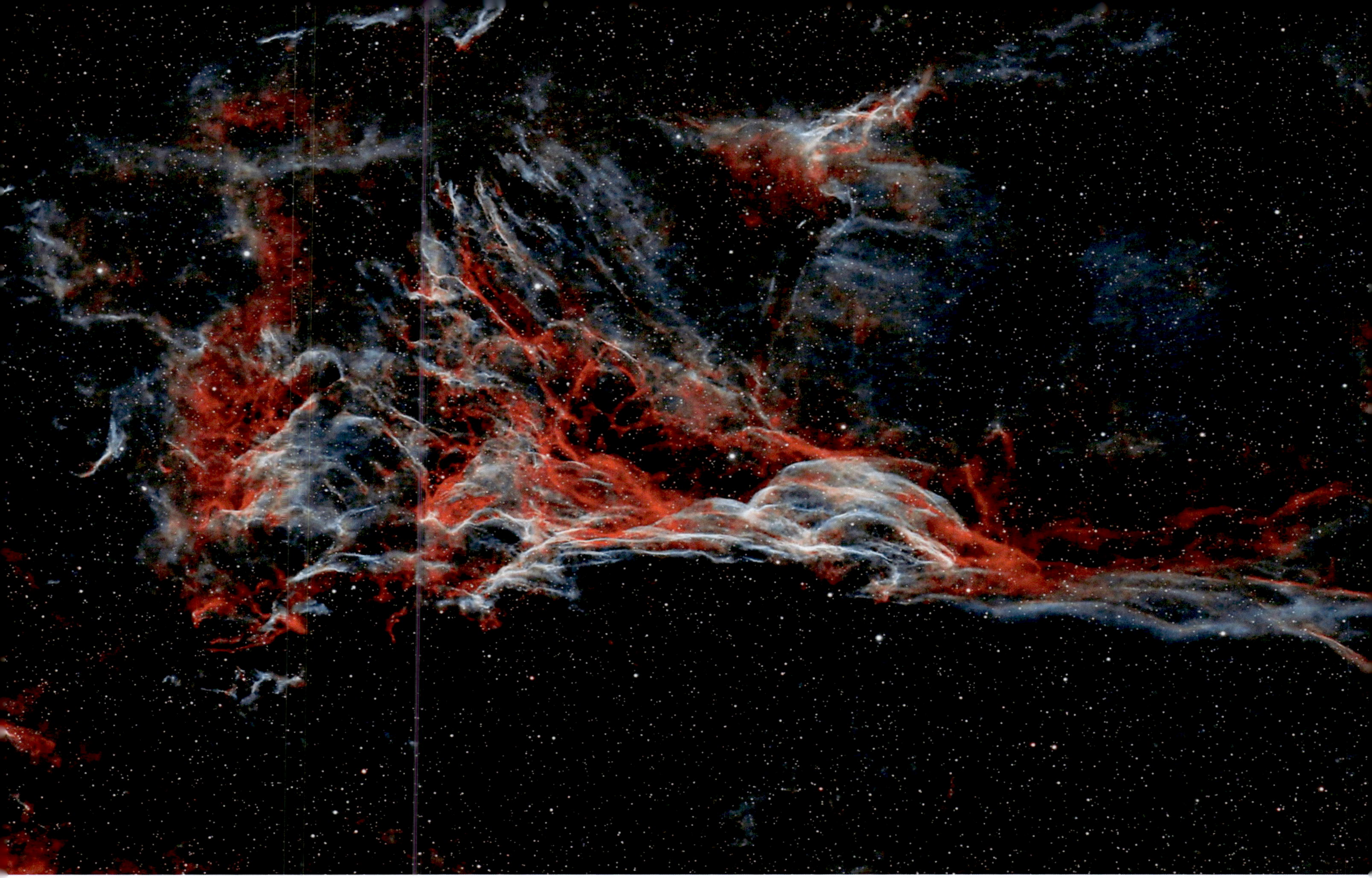

The Veil Nebula (Pickering's Triangle)
A bright part of the Cygnus Loop supernova remnant in the constellation Cygnus
Approximate distance from Earth: 2,400 light-years

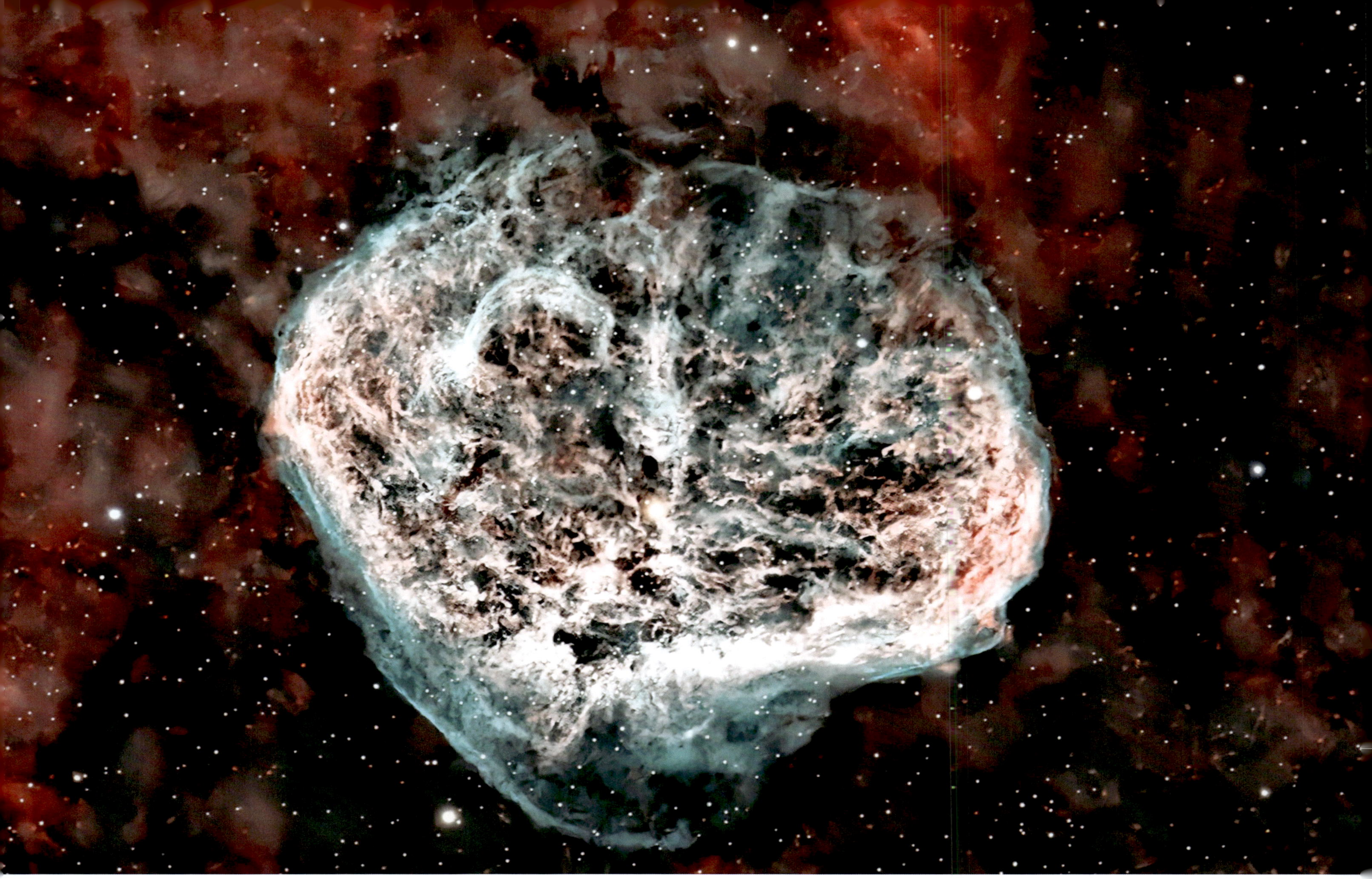

The Crescent Nebula (NGC 6888)
An emission nebula in the constellation Cygnus
Approximate distance from Earth: 5,000 light-years

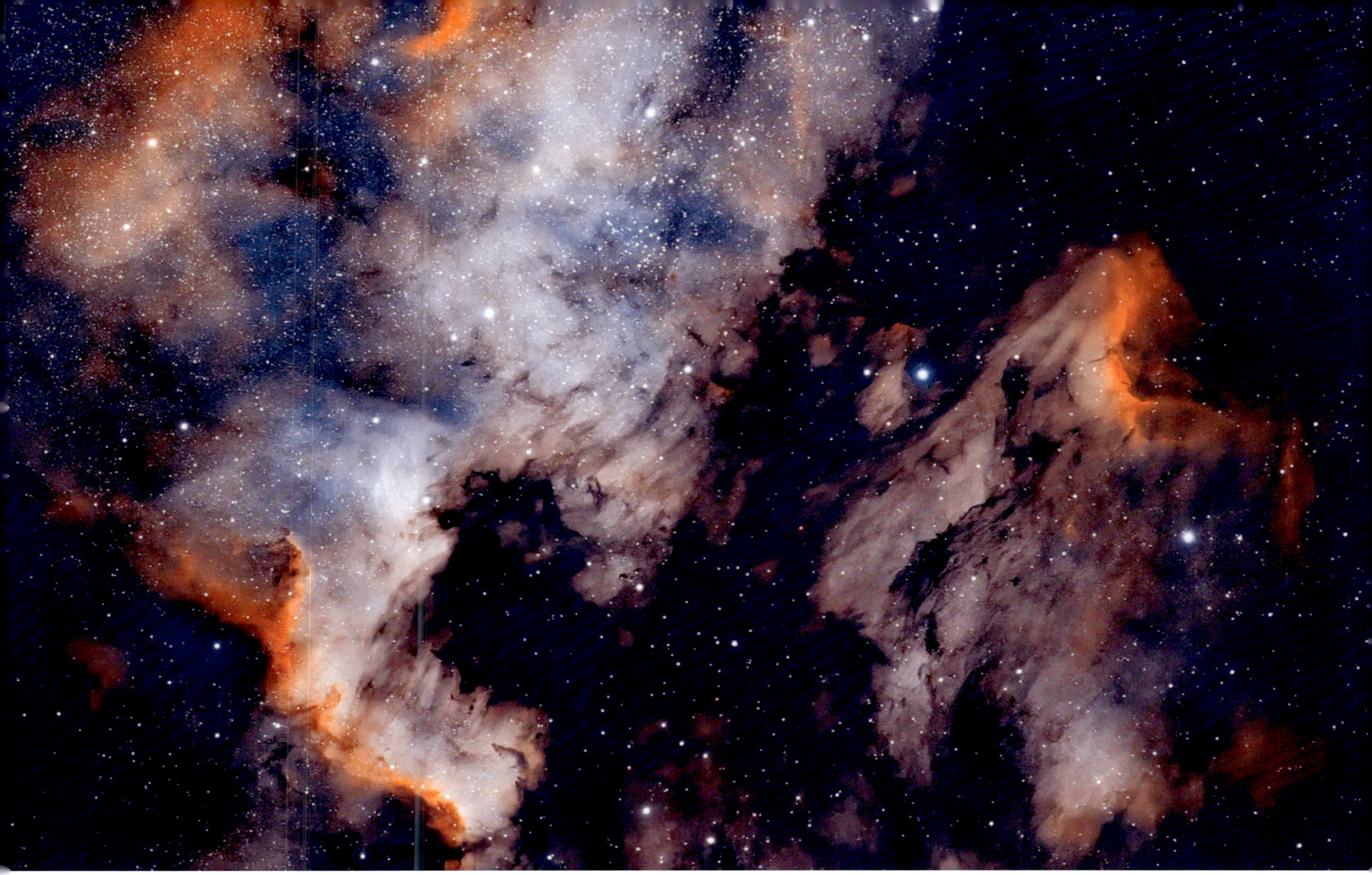

North American Nebula & Pelican Nebula — 2 panel mosaic
An interconnected region of two emission nebulae in the constellation Cygnus
Approximate distance from Earth: 1,800 light-years

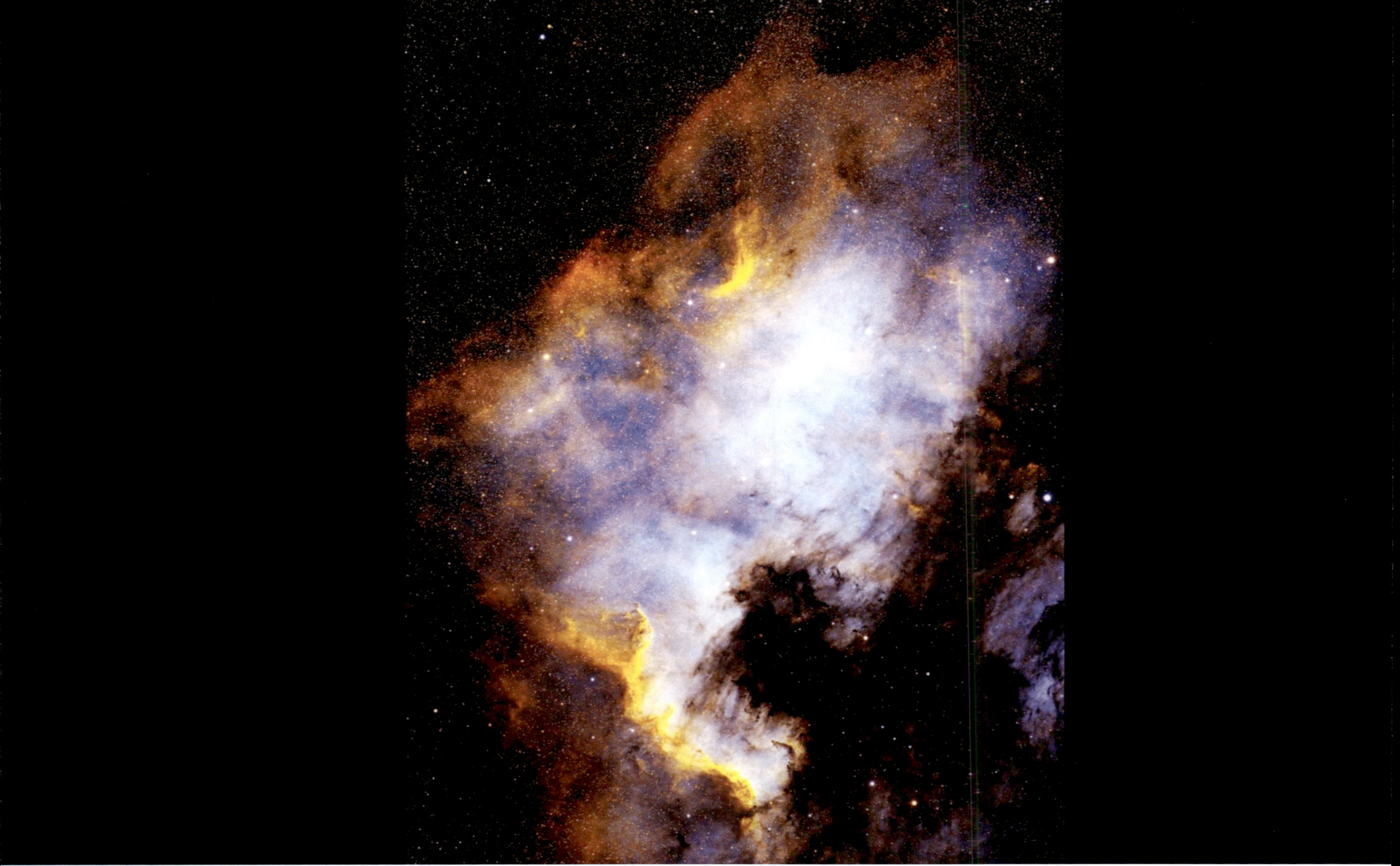

North American Nebula (NGC 7000)
An emission nebula in the constellation Cygnus
Approximate distance from Earth: 1,800 light-years

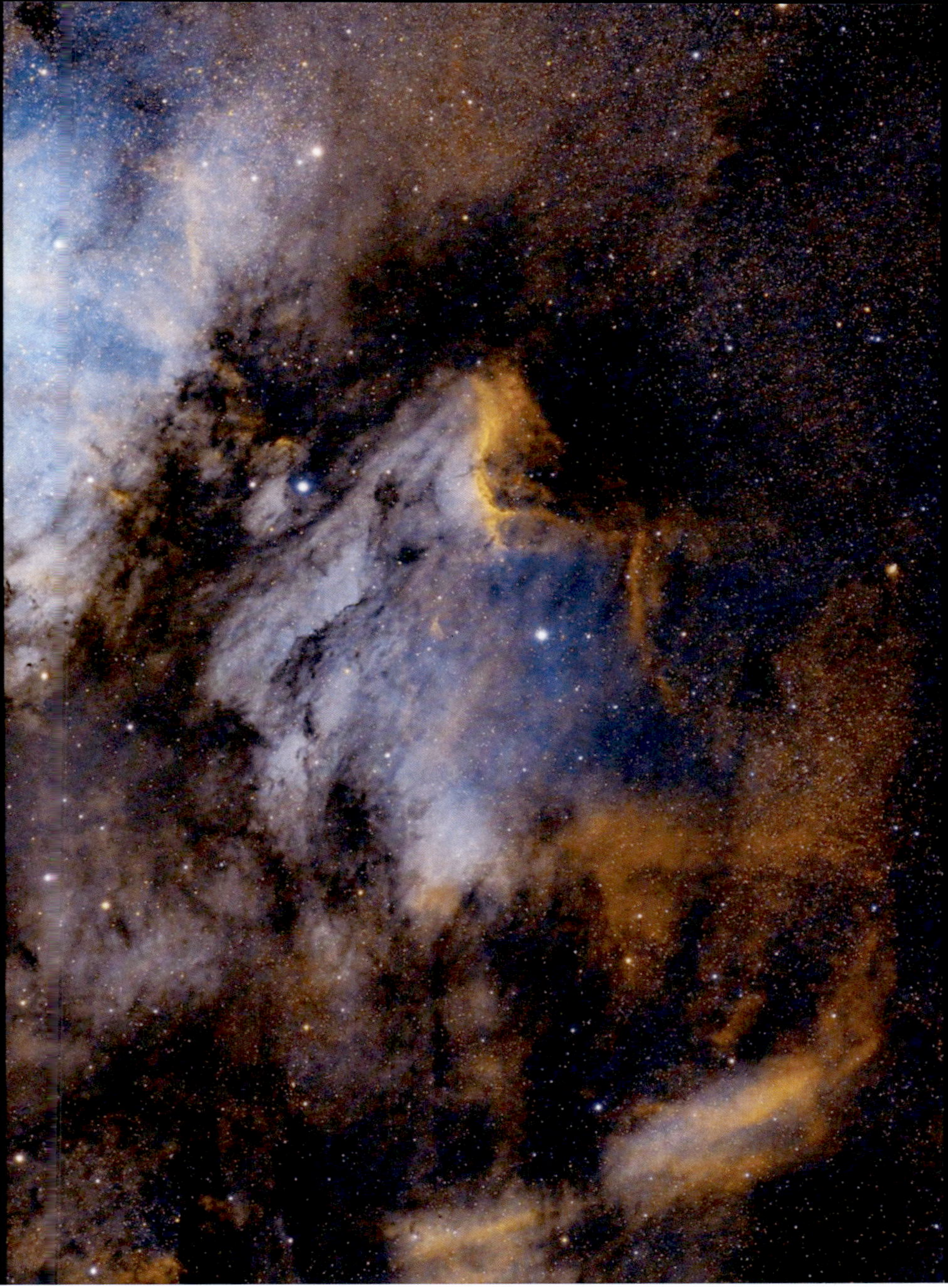

The Pelican Nebula (IC 5070)
An emission nebula in the constellation Cygnus
Approximate distance from Earth: 1,800 light-years

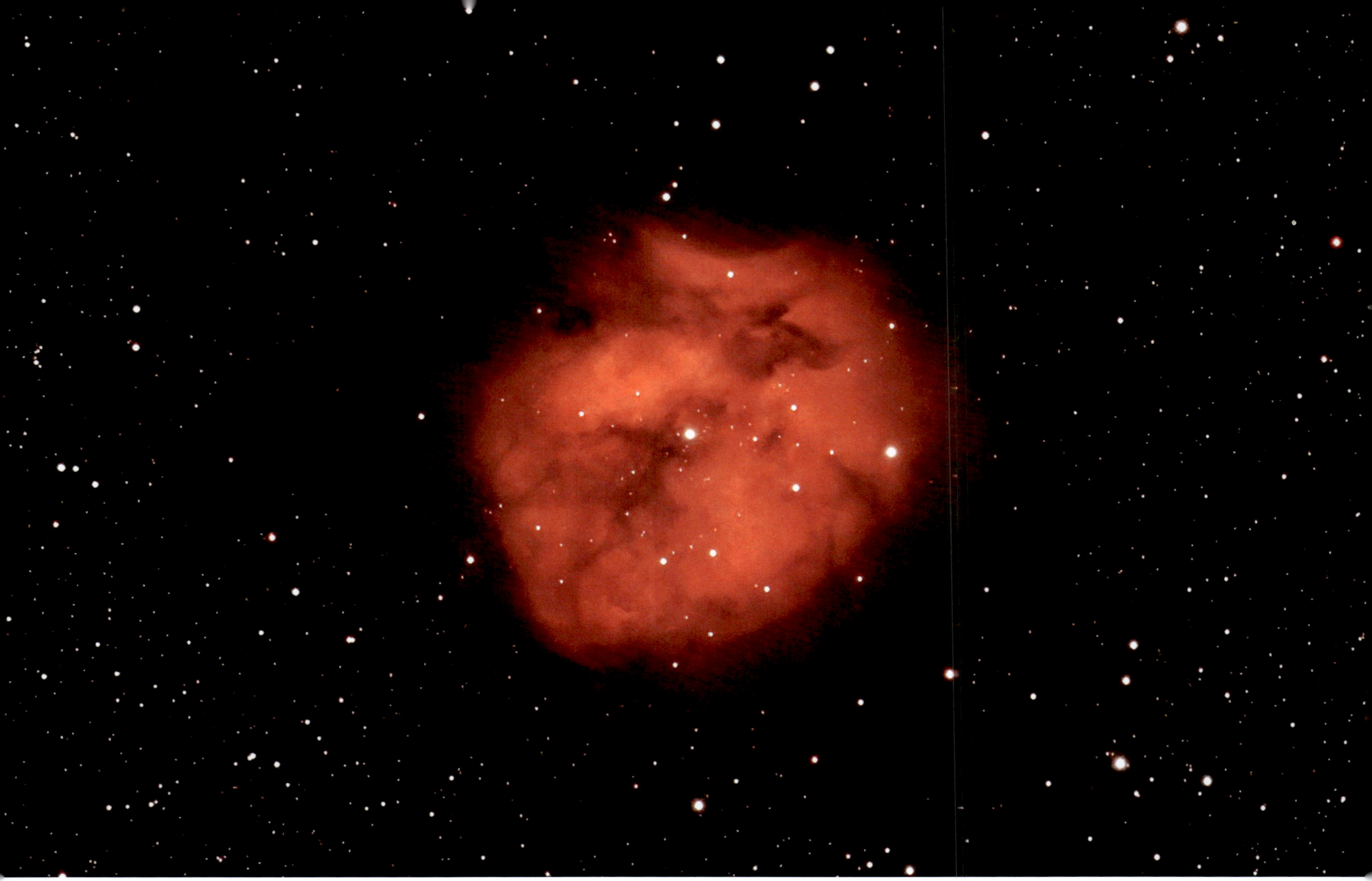

Cocoon Nebula (IC 5146)
An emission nebula (with reflection and dark-nebula components) in the constellation Cygnus
Approximate distance from Earth: 3,300 light-years

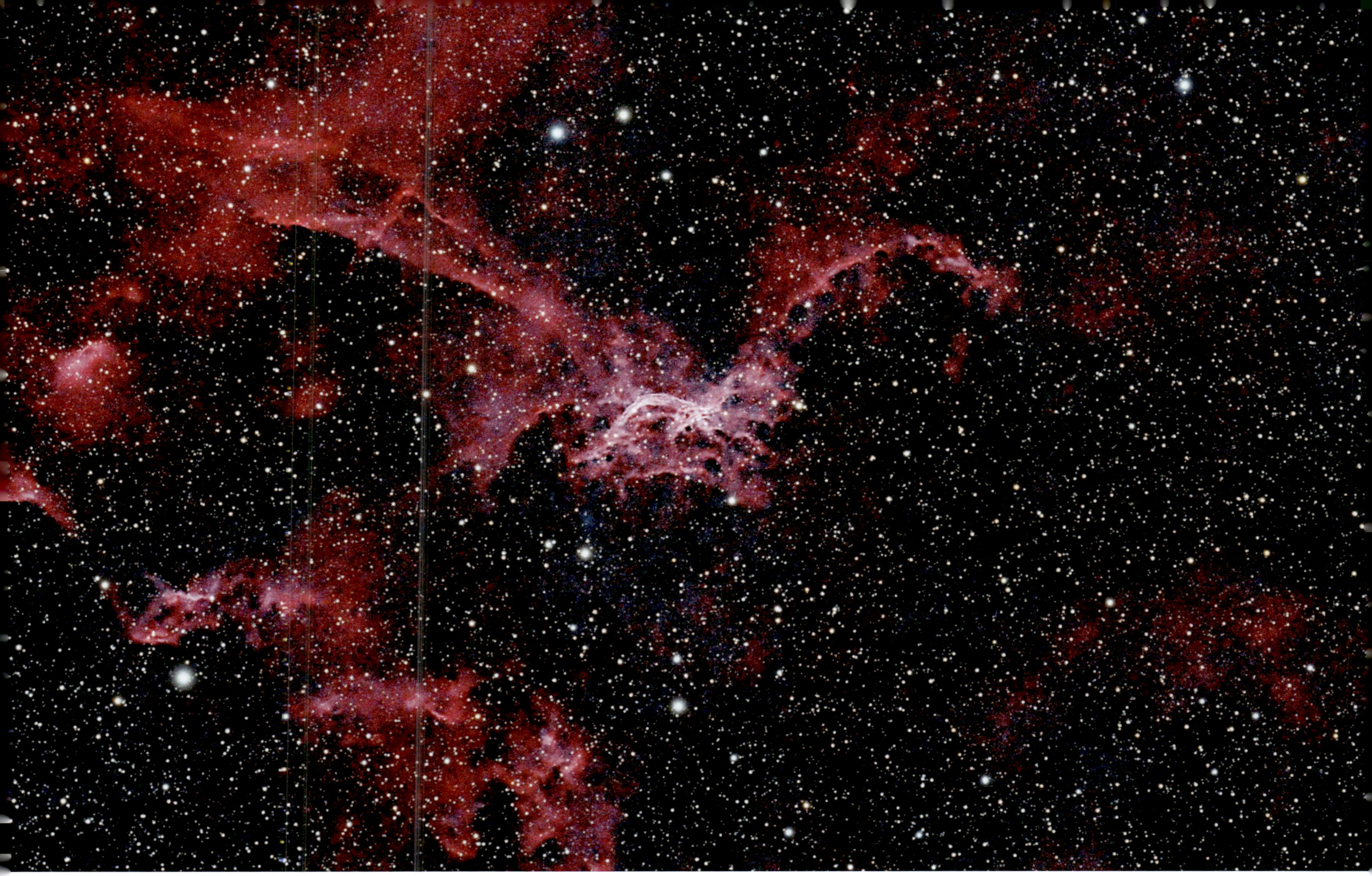

Flying Dragon Nebula (Sharpless 2-114)
A faint emission nebula in the constellation Cygnus
Approximate distance from Earth: 6,000 light-years

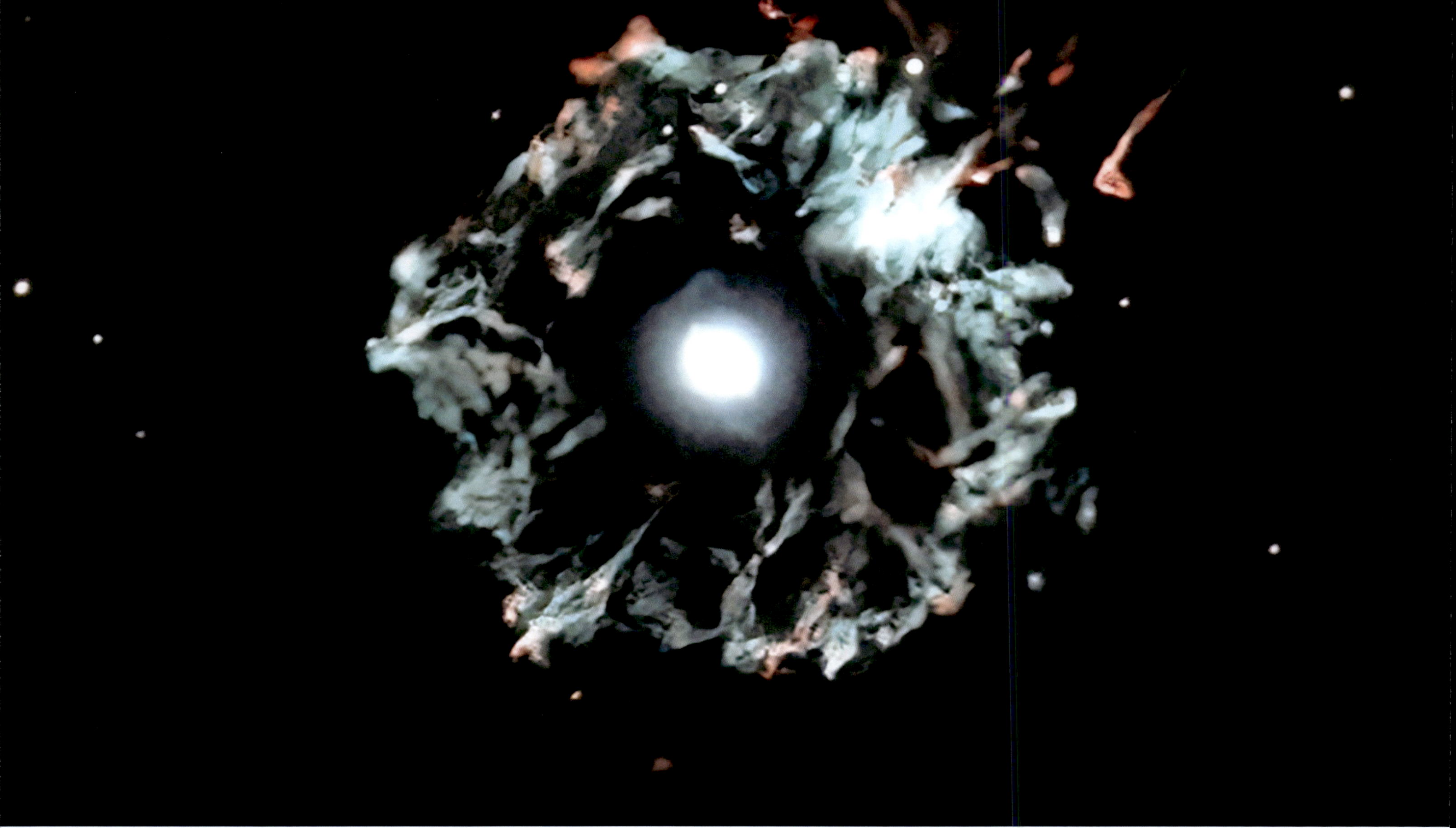

The Cat's Eye Nebula (NGC 6543)
A planetary nebula in the constellation Draco
Approximate distance from Earth: 3,300 light-years

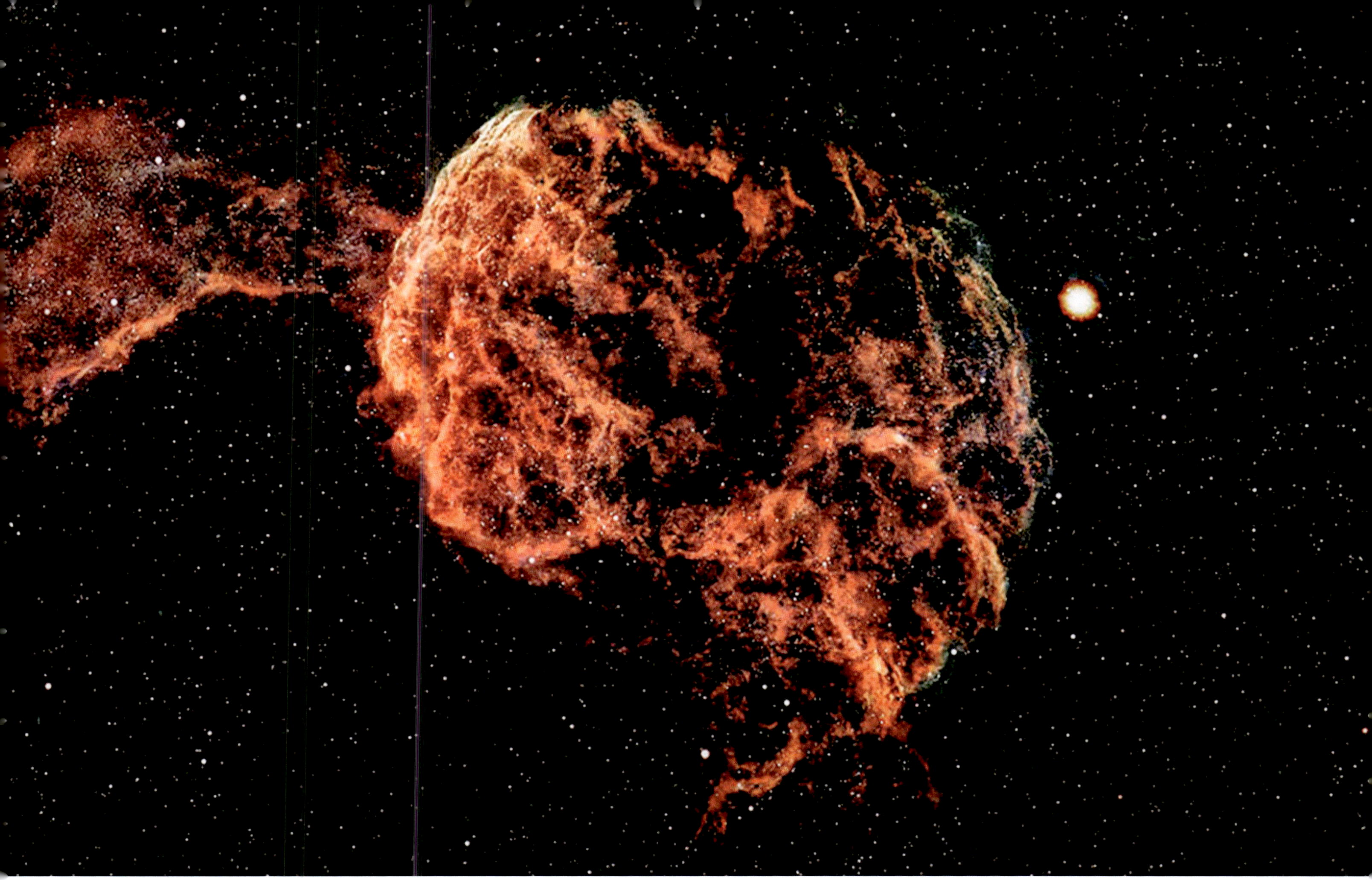

The Jelly Fish Nebula (IC 443)
A supernova remnant in the constellation Gemini
Approximate distance from Earth: 5,000 light-years

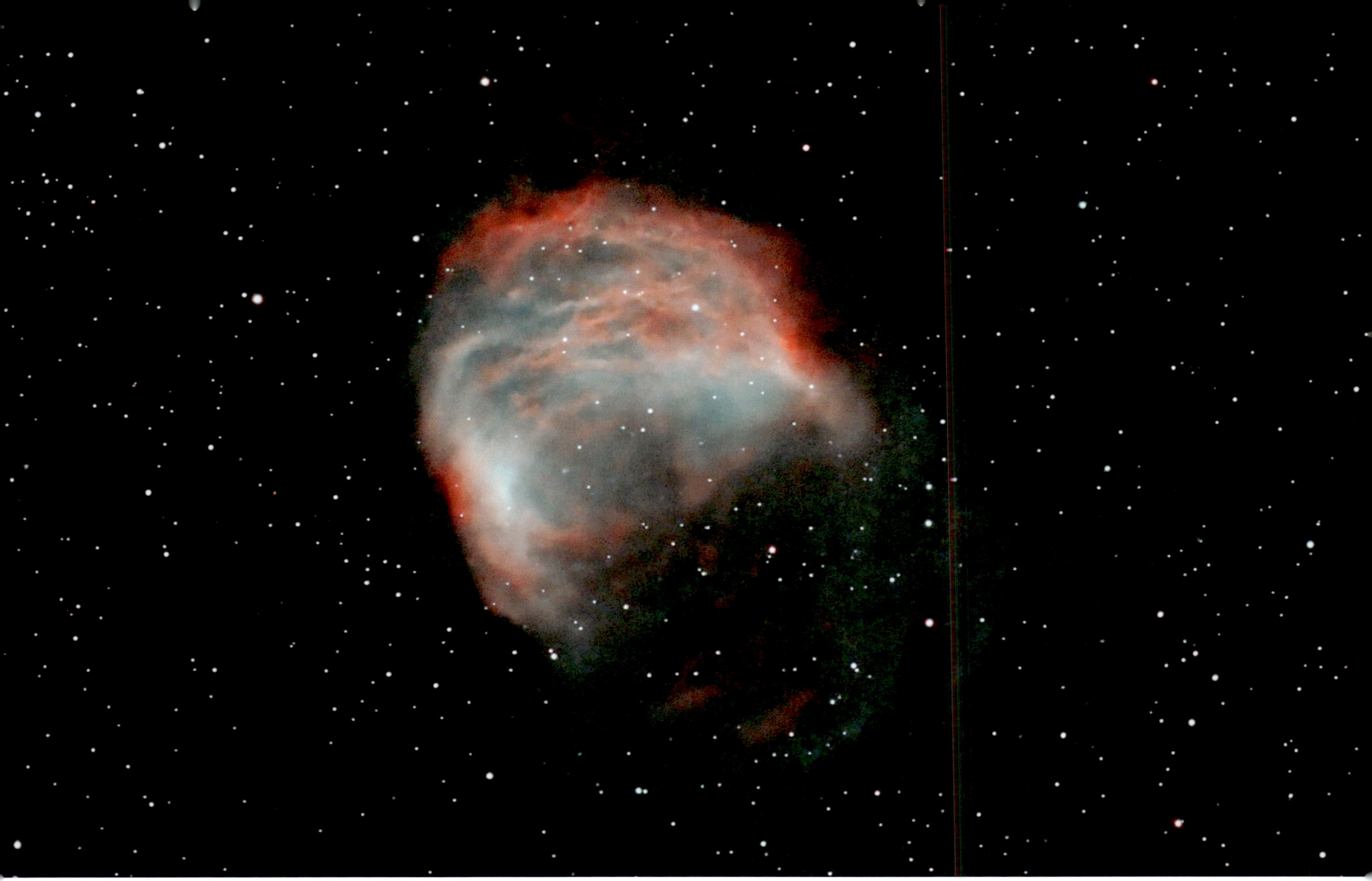

Medusa Nebula (Sh2-274)
A planetary nebula in the constellation Gemini
Approximate distance from Earth: 1,500 light-years

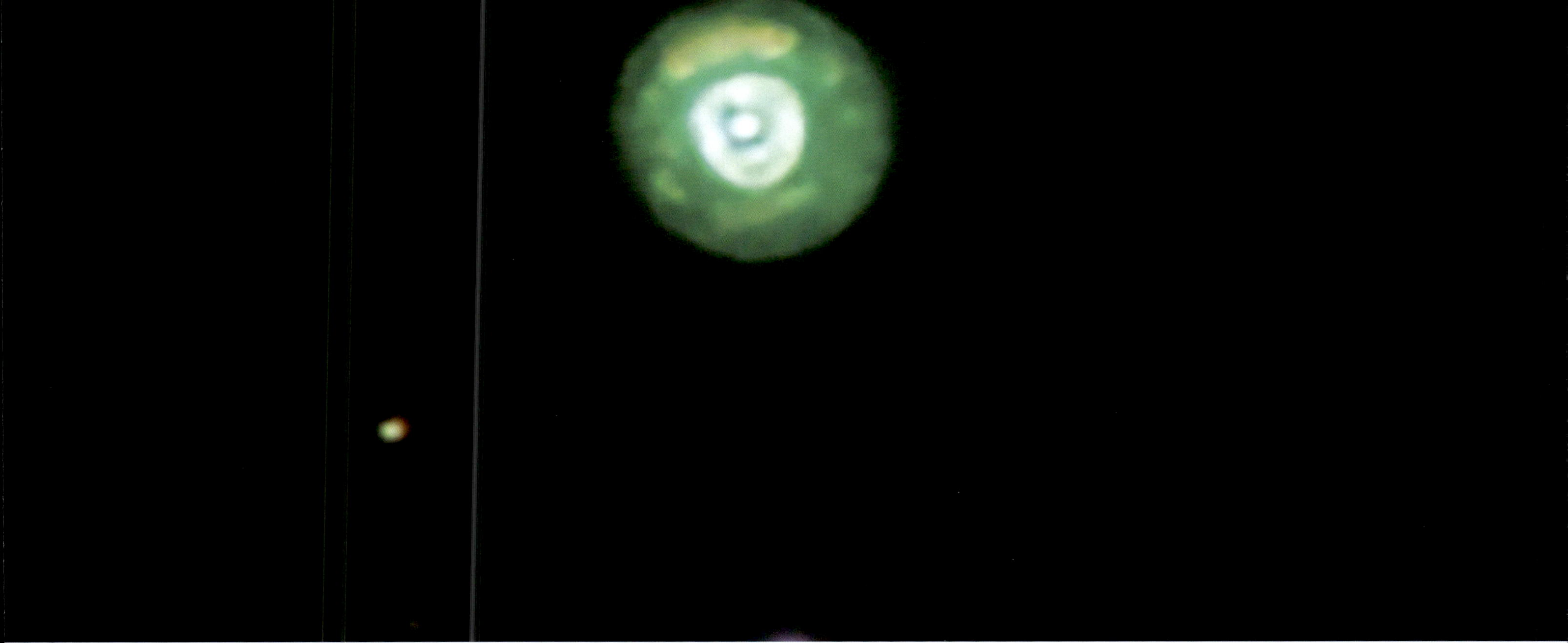

The Eskimo Nebula (NGC 2392)
A planetary nebula in the constellation Gemini
Approximate distance from Earth: 2,900 light-years

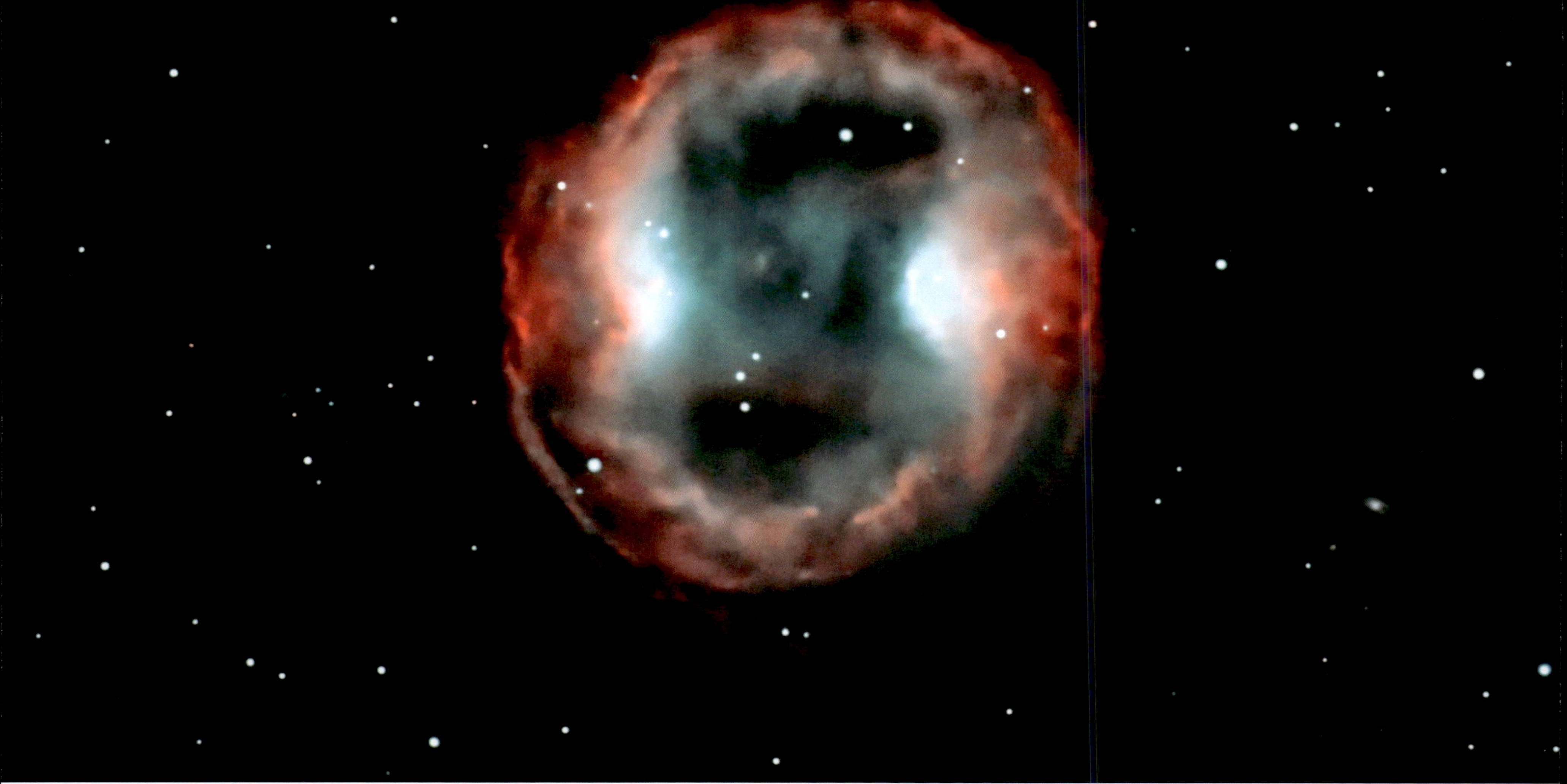

Headphone Nebula (PK 164+31.1)
A planetary nebula in the constellation Lynx
Approximate distance from Earth: 1,600 light-years

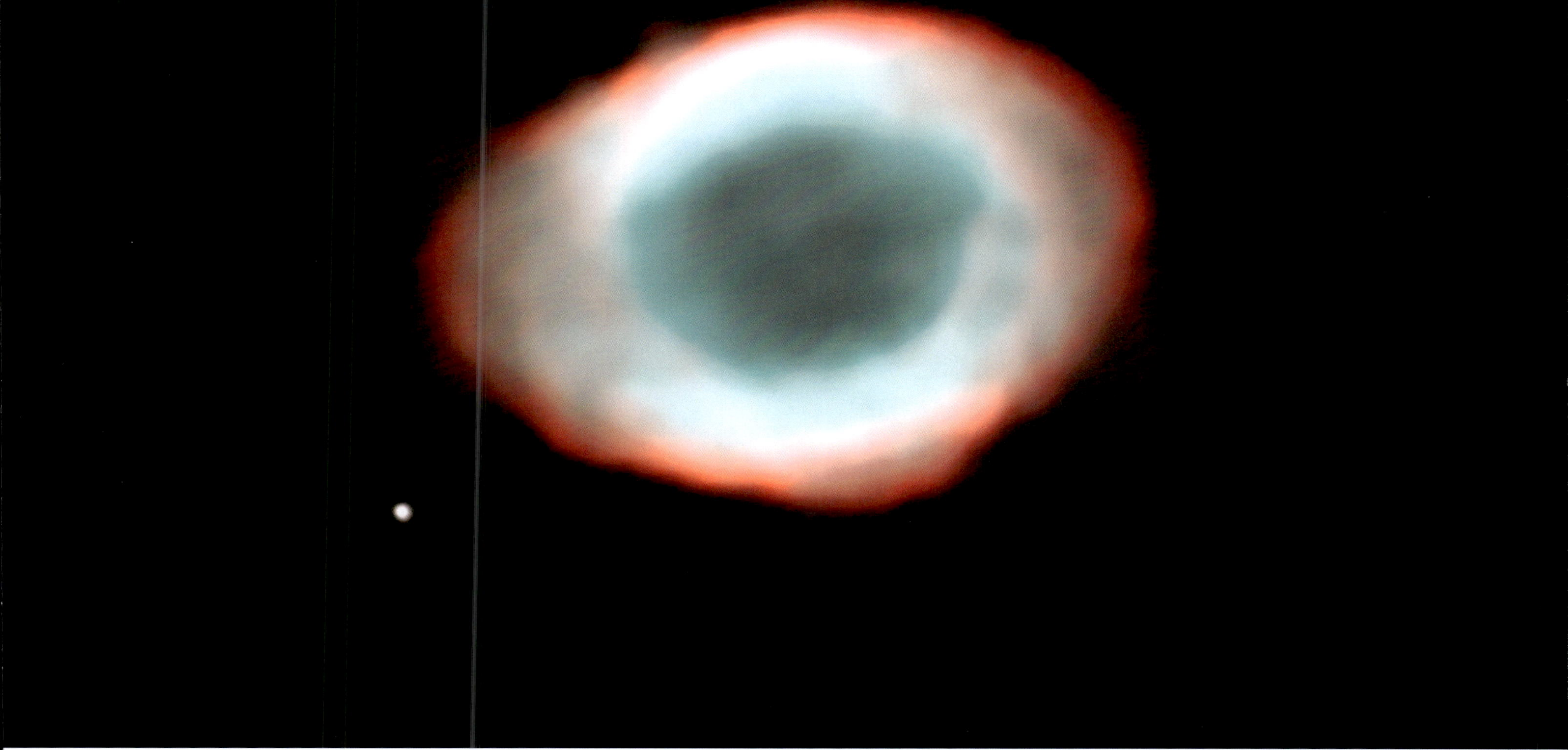

Messier 57 (The Ring Nebula)
A planetary nebula in the constellation Lyra
Approximate distance from Earth: 2,300 light-years

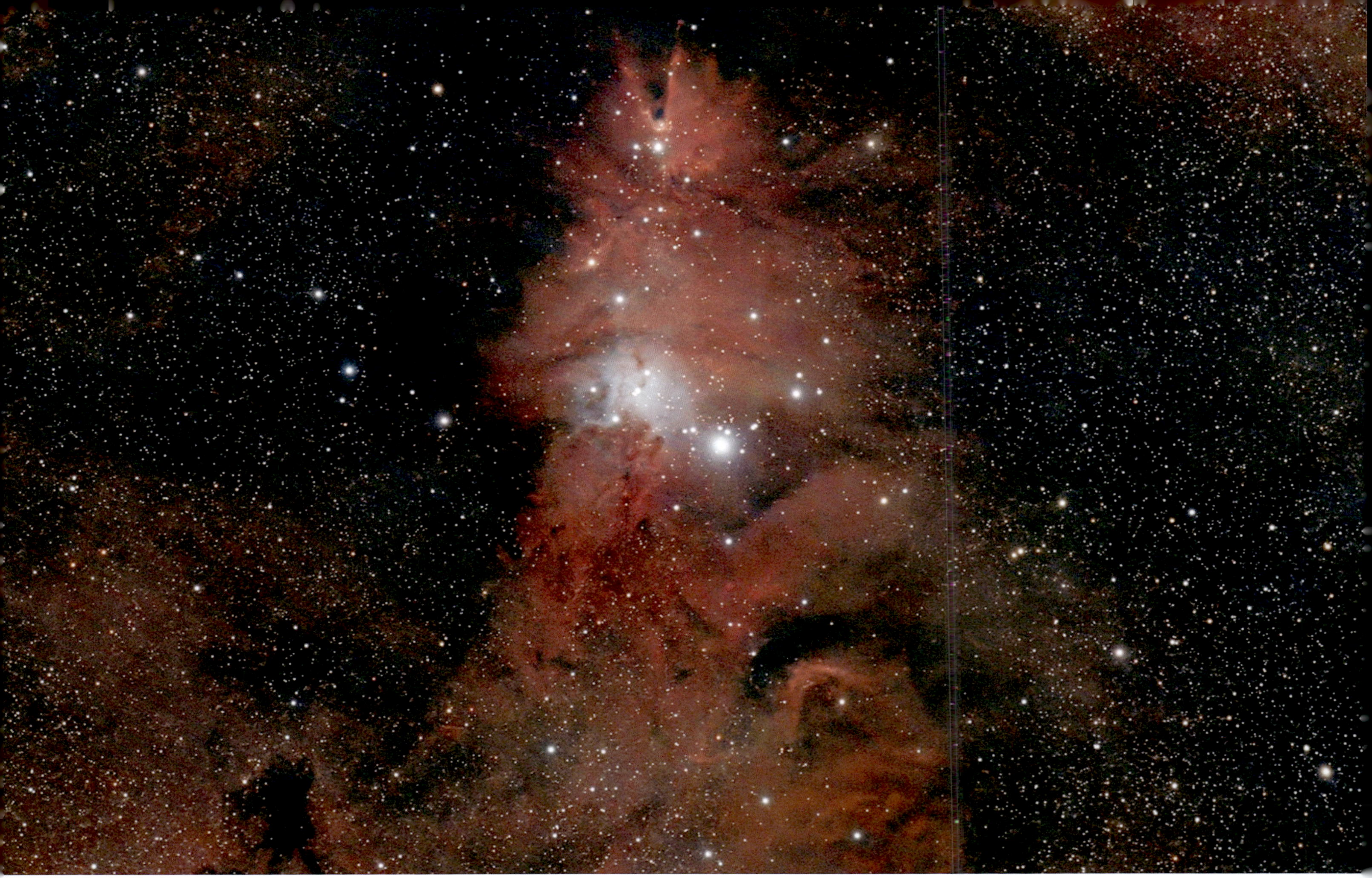

Christmas Tree Nebula / Cluster (NGC 2264)
An emission nebula and a star cluster in the constellation Monoceros
Approximate distance from Earth: 2,400 light-years

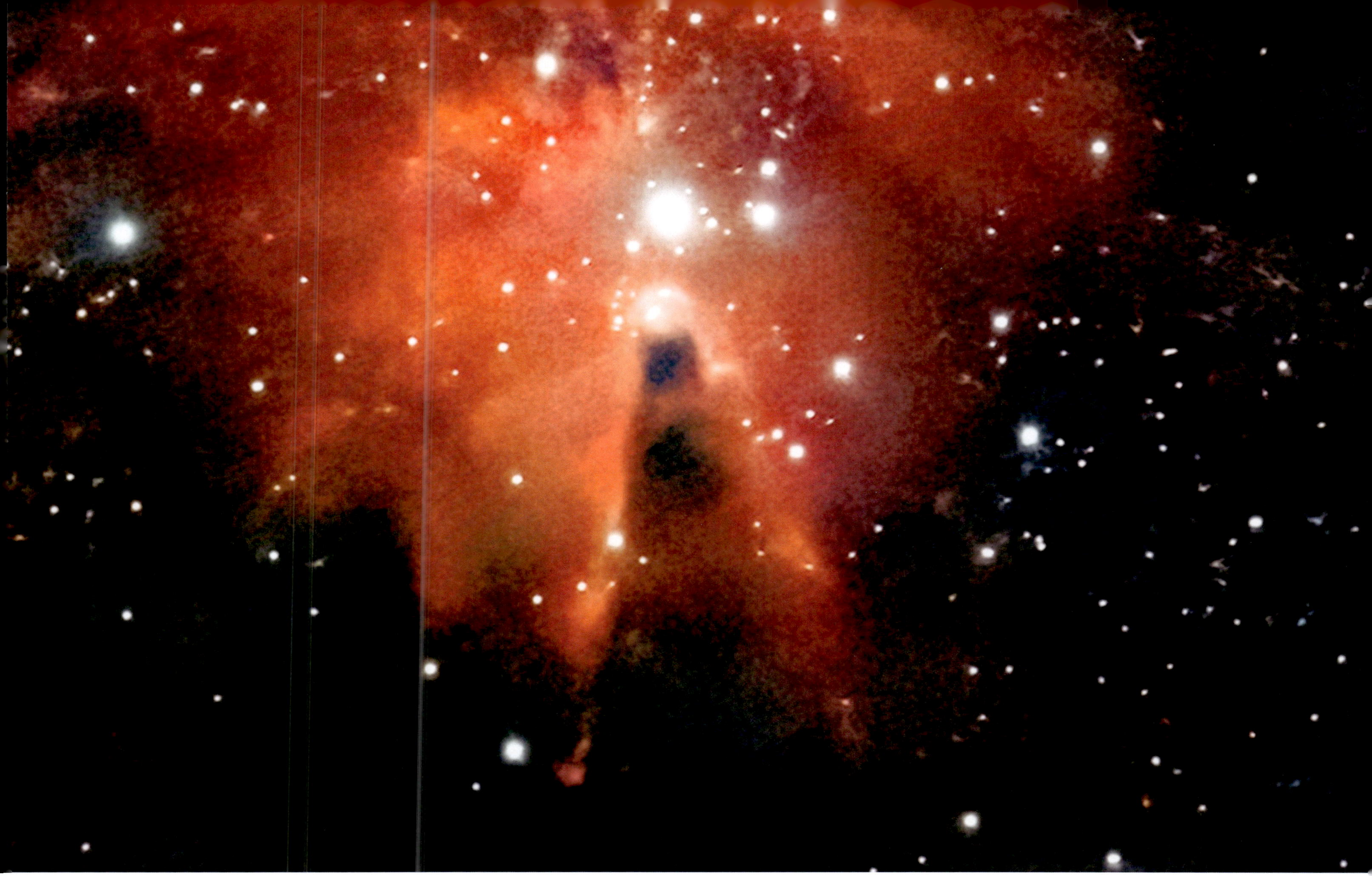

The Cone Nebula (NGC 2264)
A dark absorption nebula in the constellation Monoceros
Approximate distance from Earth: 2,400 light-years

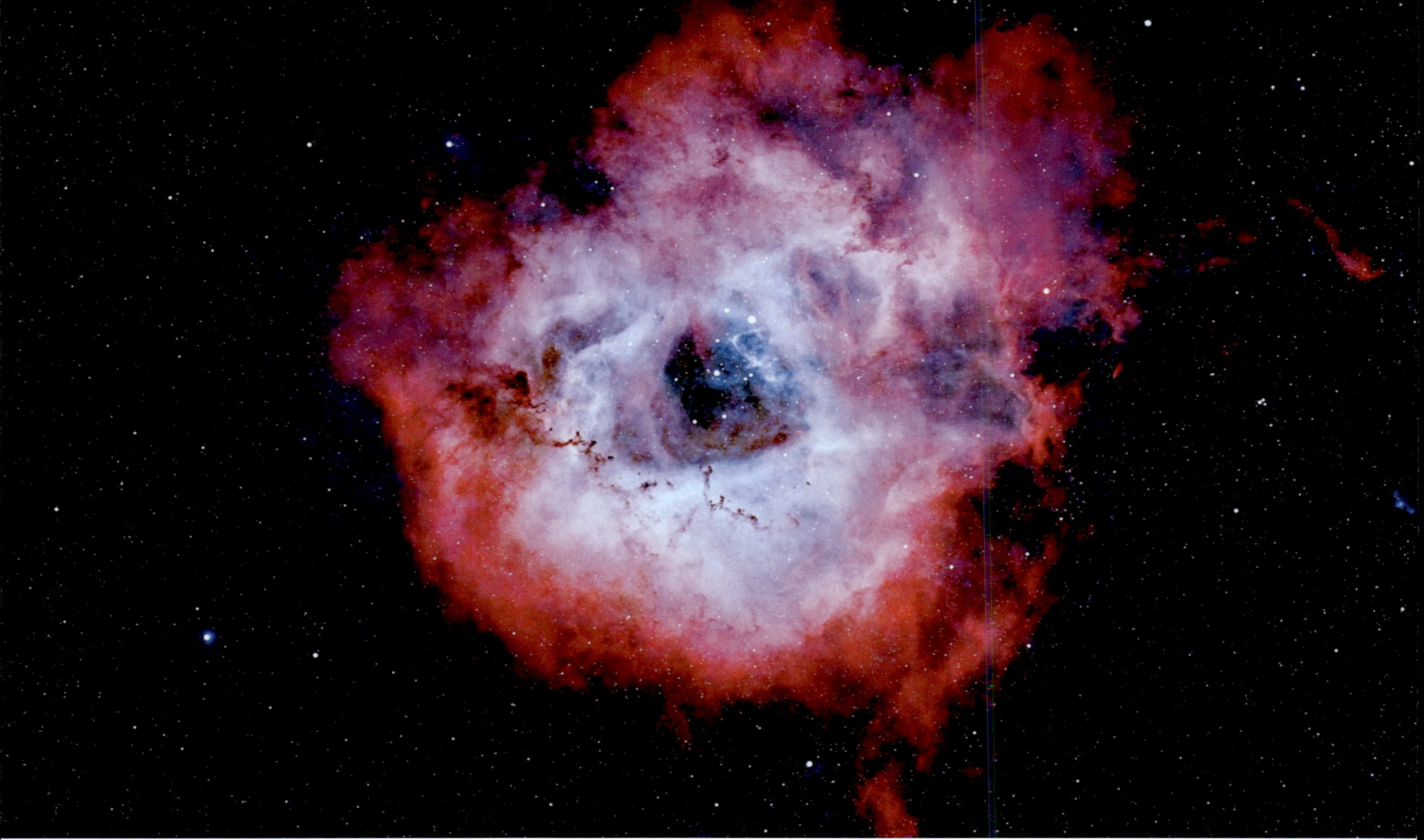

The Rosette Nebula (NGC 2237)
An emission nebula in the constellation Monoceros
Approximate distance from Earth: 5,200 light-years

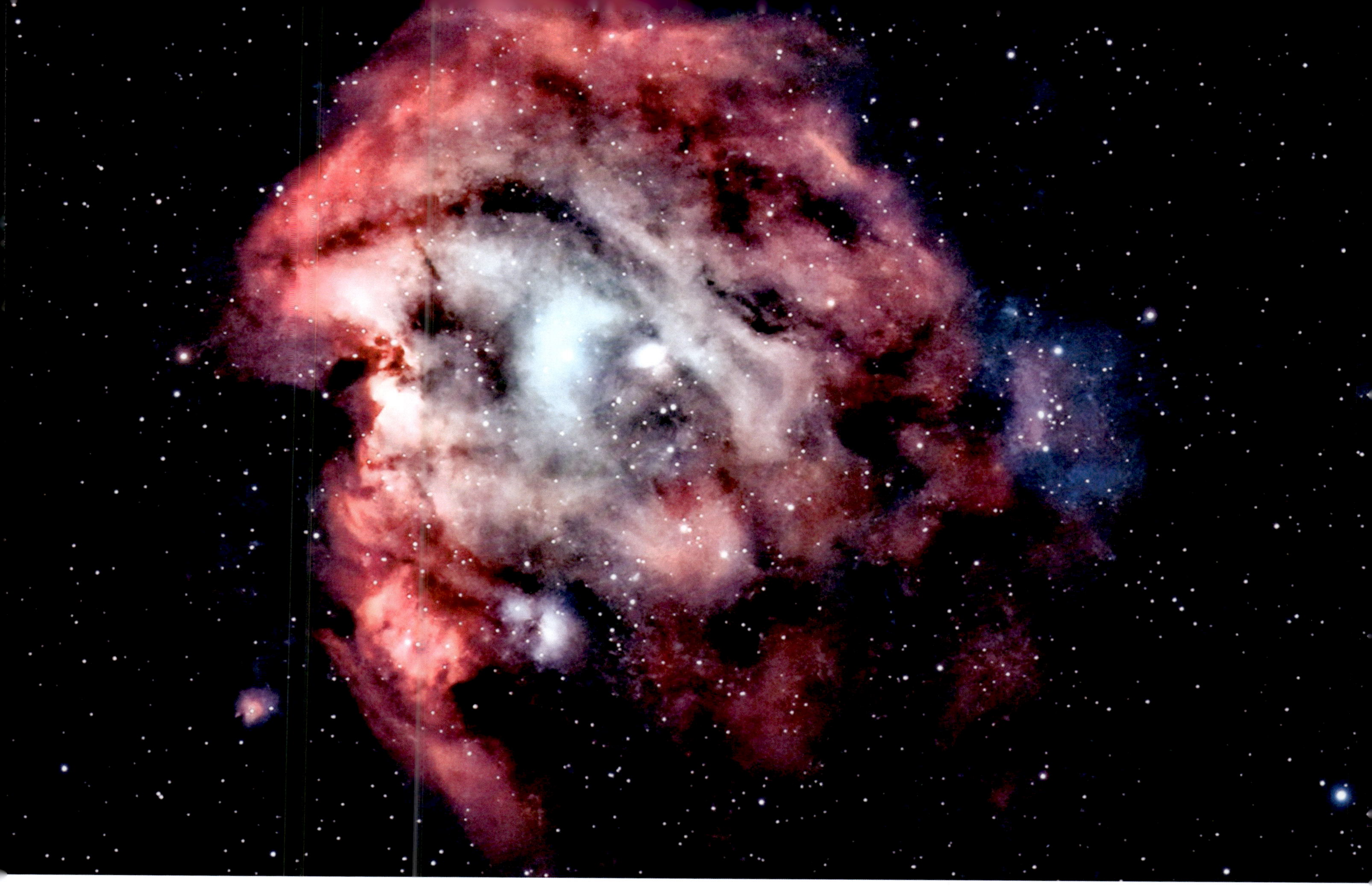

The Monkey Head Nebula (NGC 2174)
An emission nebula in the constellation Orion
Approximate distance from Earth: 6,400 light-years

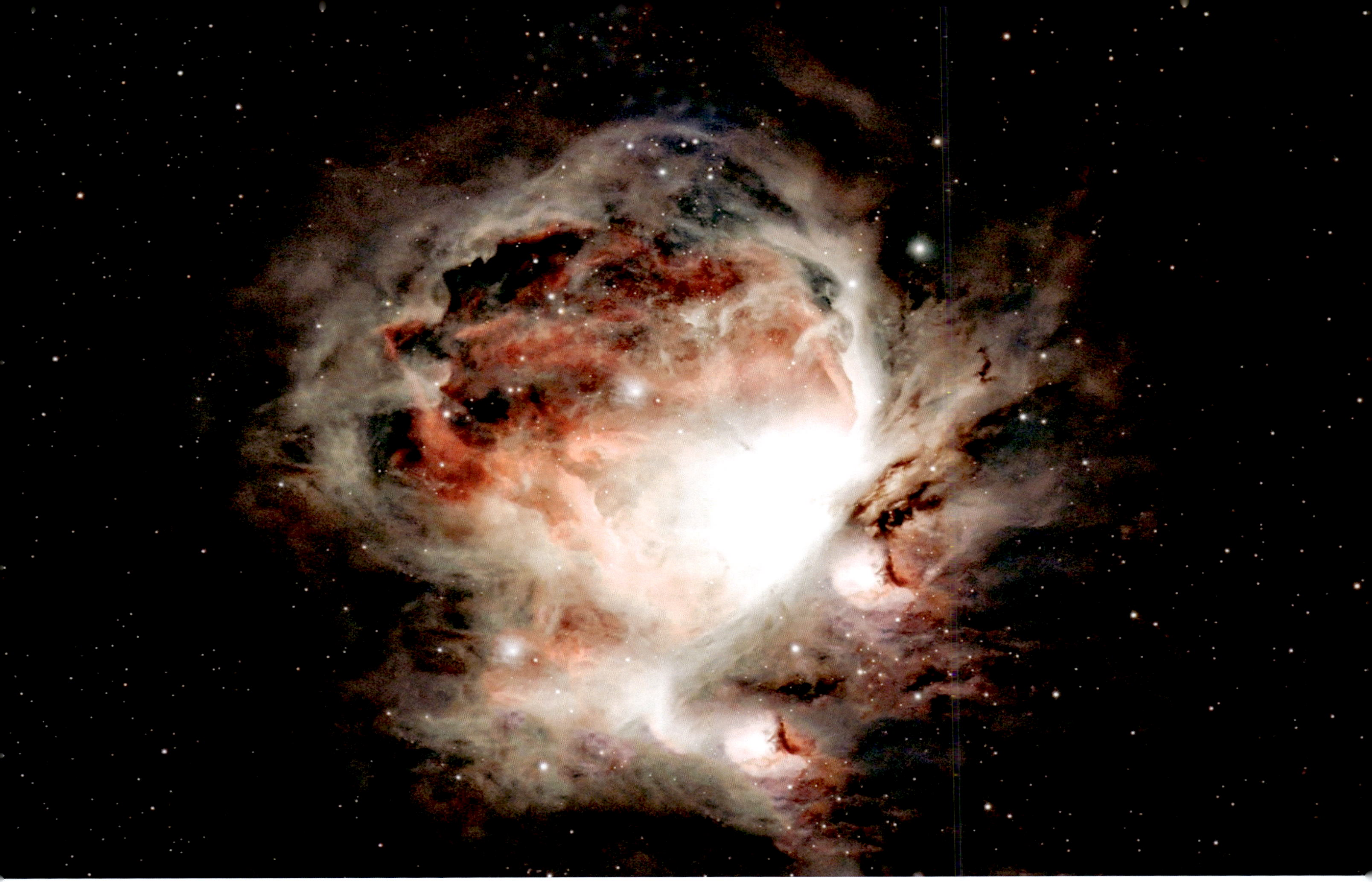

Messier 42 (The Orion Nebula)
A diffuse emission nebula / stellar nursery in the constellation Orion
Approximate distance from Earth: 1,350 light-years

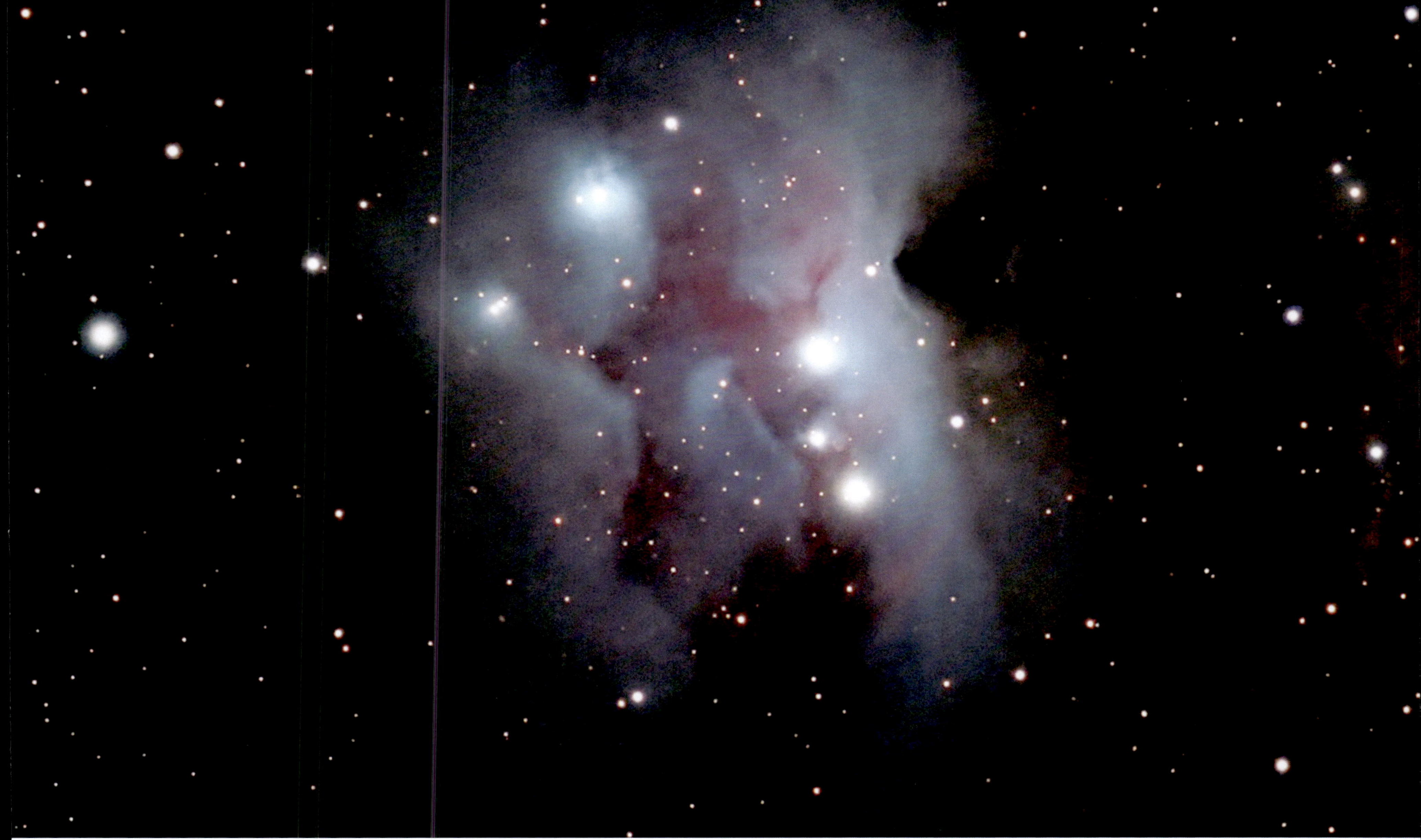

Running Man Nebula (NGC 1977)
A reflection nebula in the constellation Orion
Approximate distance from Earth: 1,500 light-years

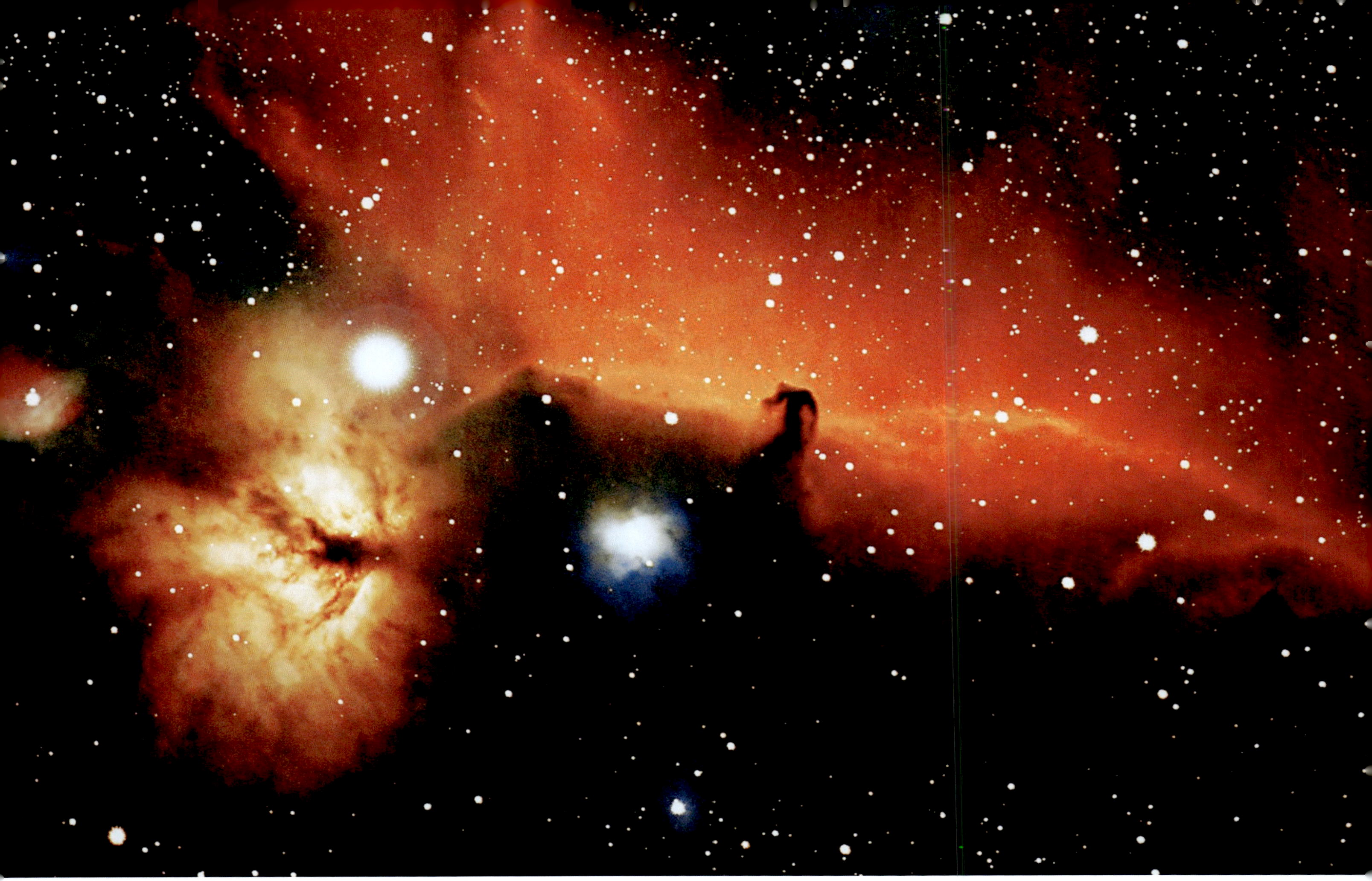

The Horsehead Nebula (B33) & Flame Nebula (NGC 2024)
B33 is a dark nebula and NGC 2024 is an emission nebula in the constellation Orion
Approximate distance from Earth: 1,400 light-years

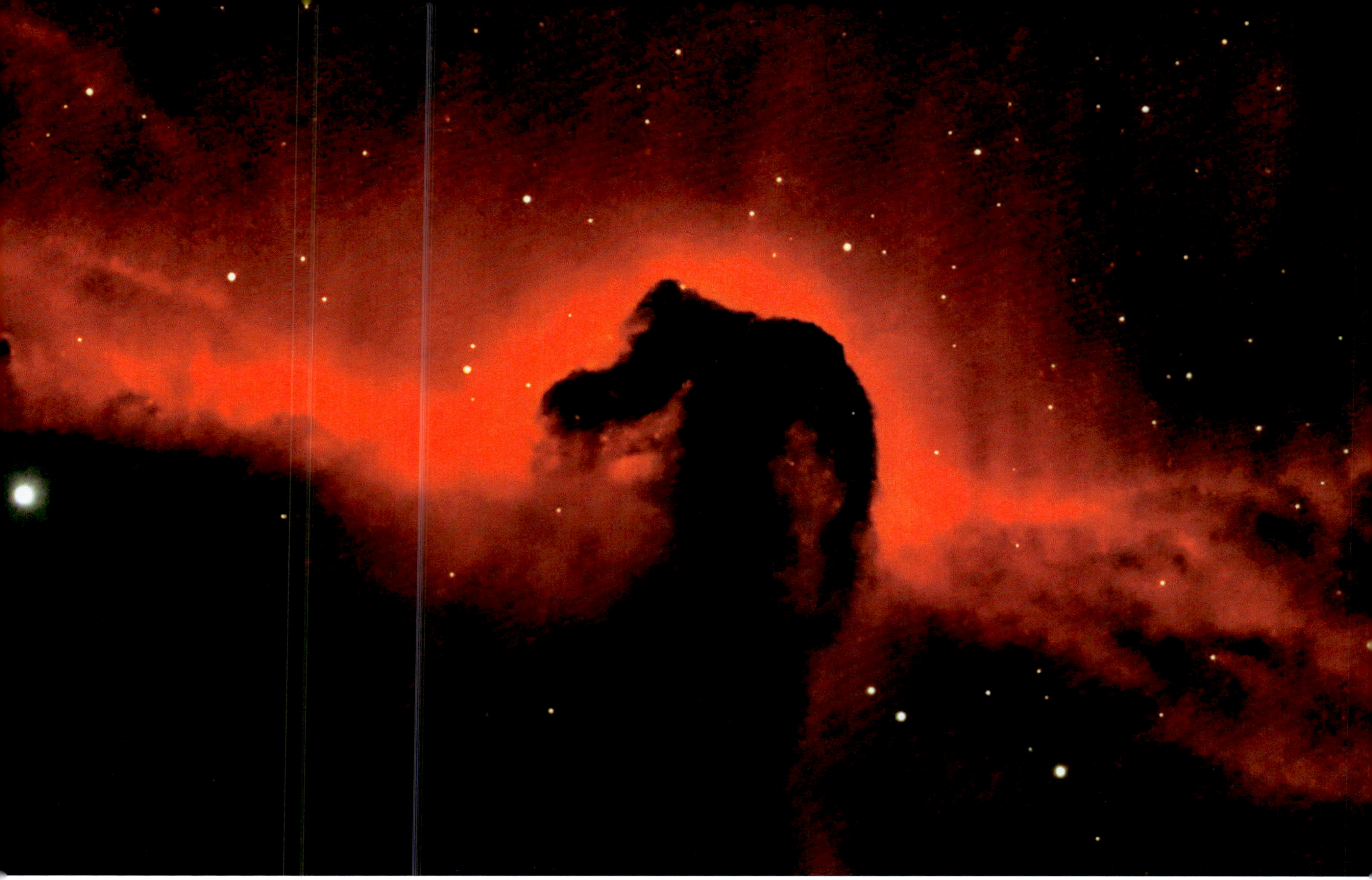

The Horsehead Nebula (B33)
A dark nebula in the constellation Orion
Approximate distance from Earth: 1,400 light-years

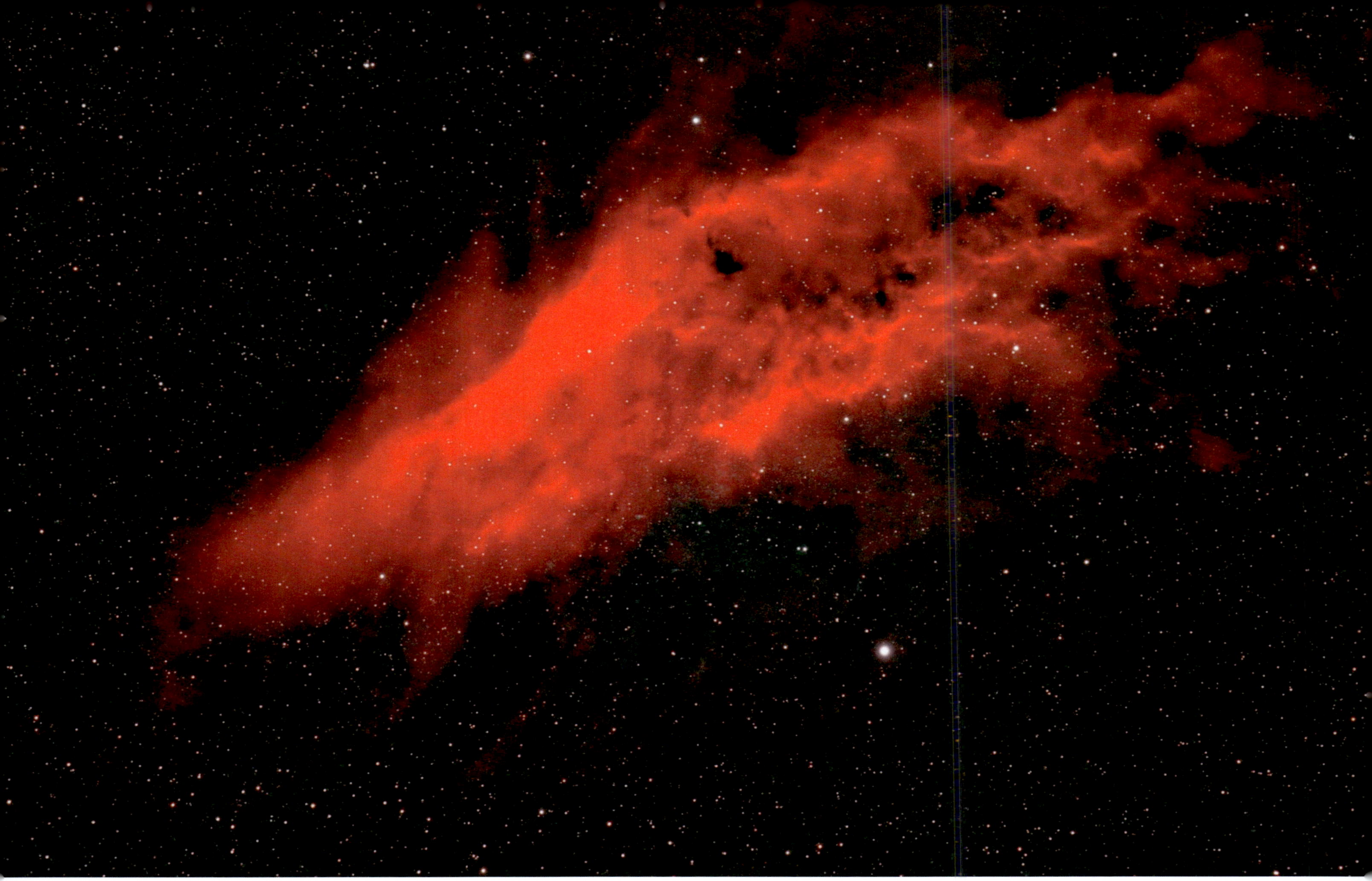

The California Nebula (NGC 1499)
An emission nebula in the constellation Perseus
Approximate distance from Earth: 1,000 light-years

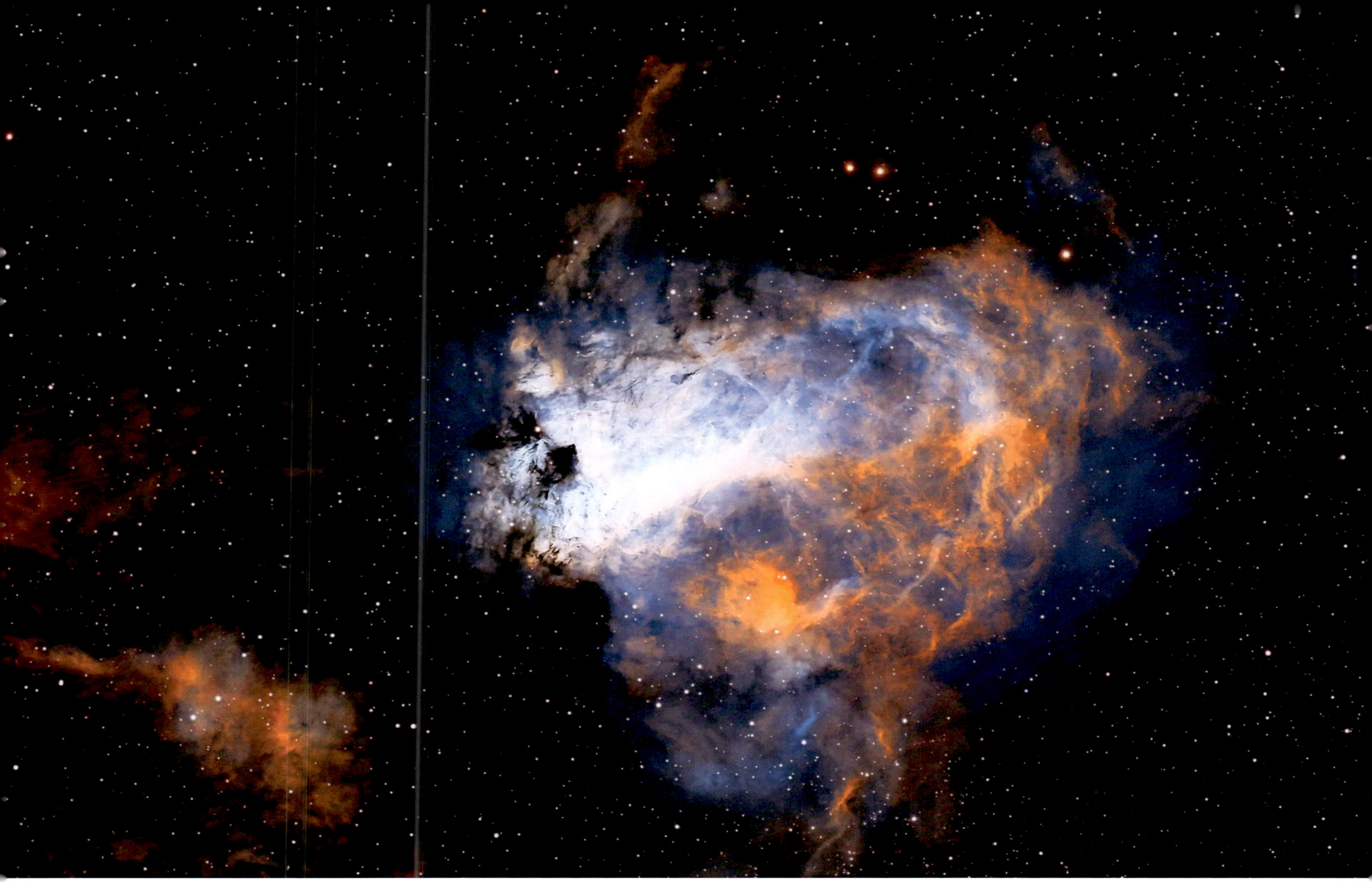

Messier 17 (The Swan Nebula)
An emission nebula and open cluster in the constellation Sagittarius
Approximate distance from Earth: 5,500 light-years

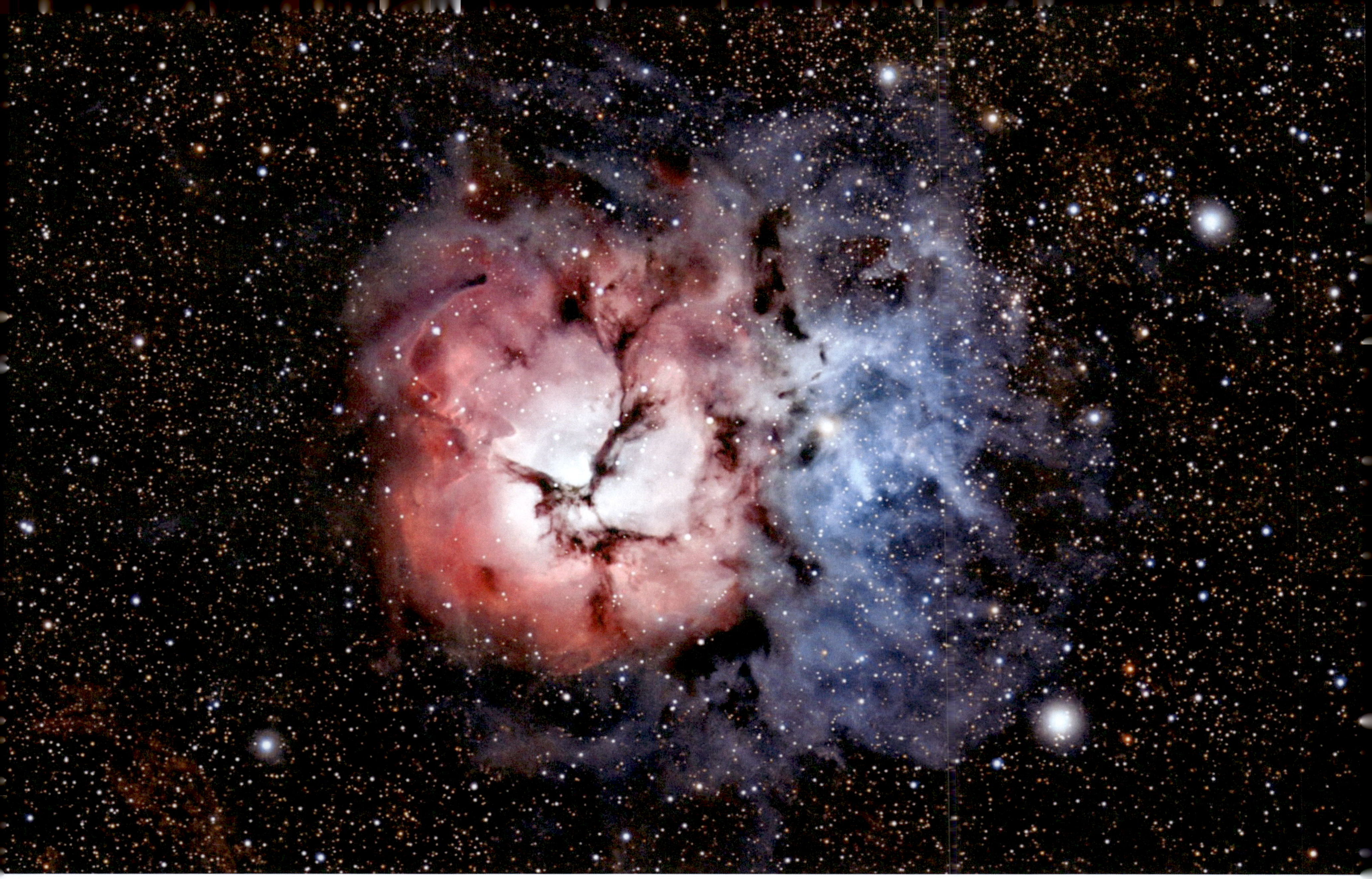

Messier 20 (The Trifid Nebula)
An emission nebula (with reflection and dark-nebula components) in the constellation Sagittarius
Approximate distance from Earth: 5,200 light-years

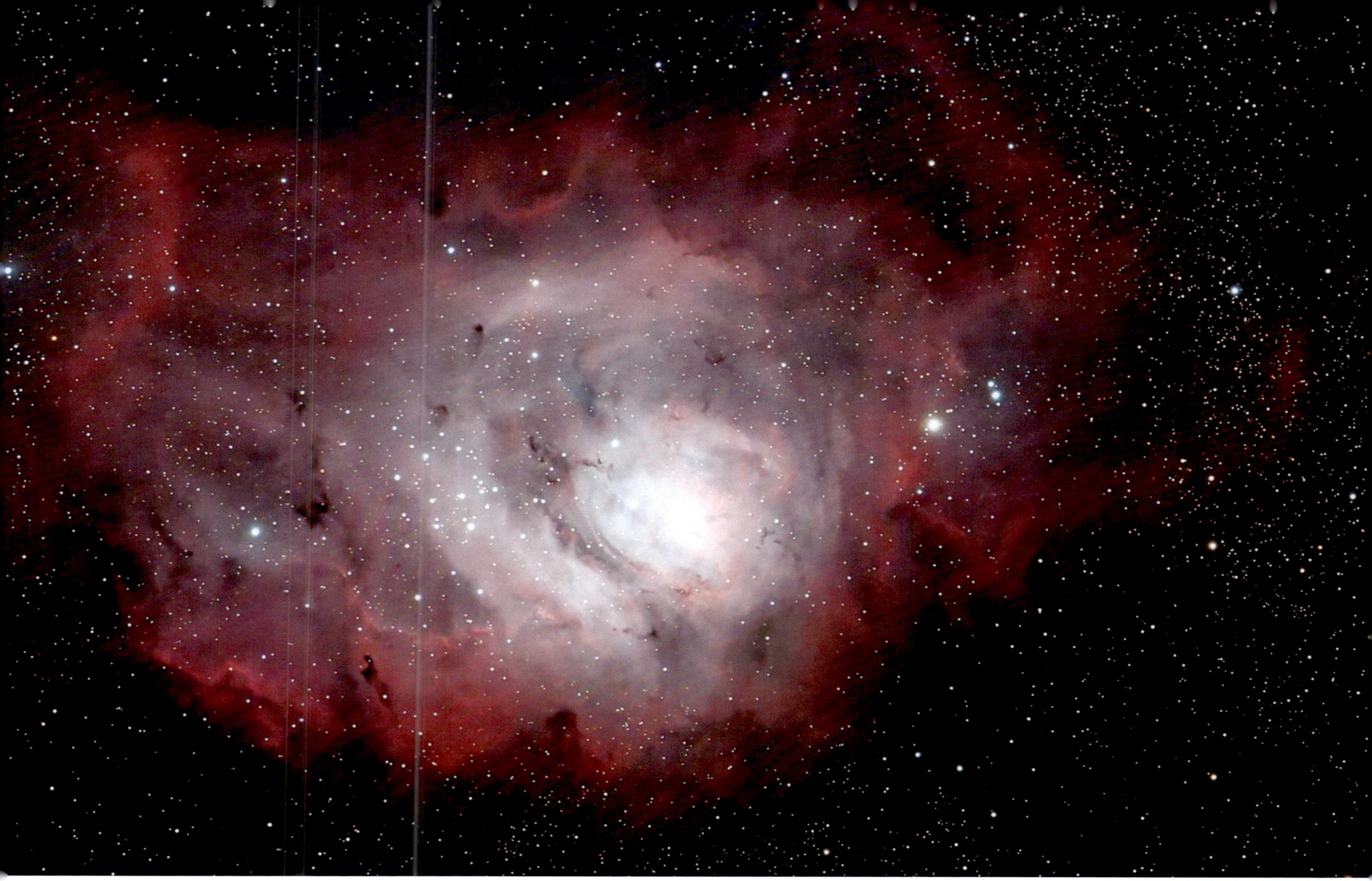

Messier 8 (The Lagoon Nebula)
An emission nebula (with areas of dark-nebula) in the constellation Sagittarius
Approximate distance from Earth: 4,100 light-years

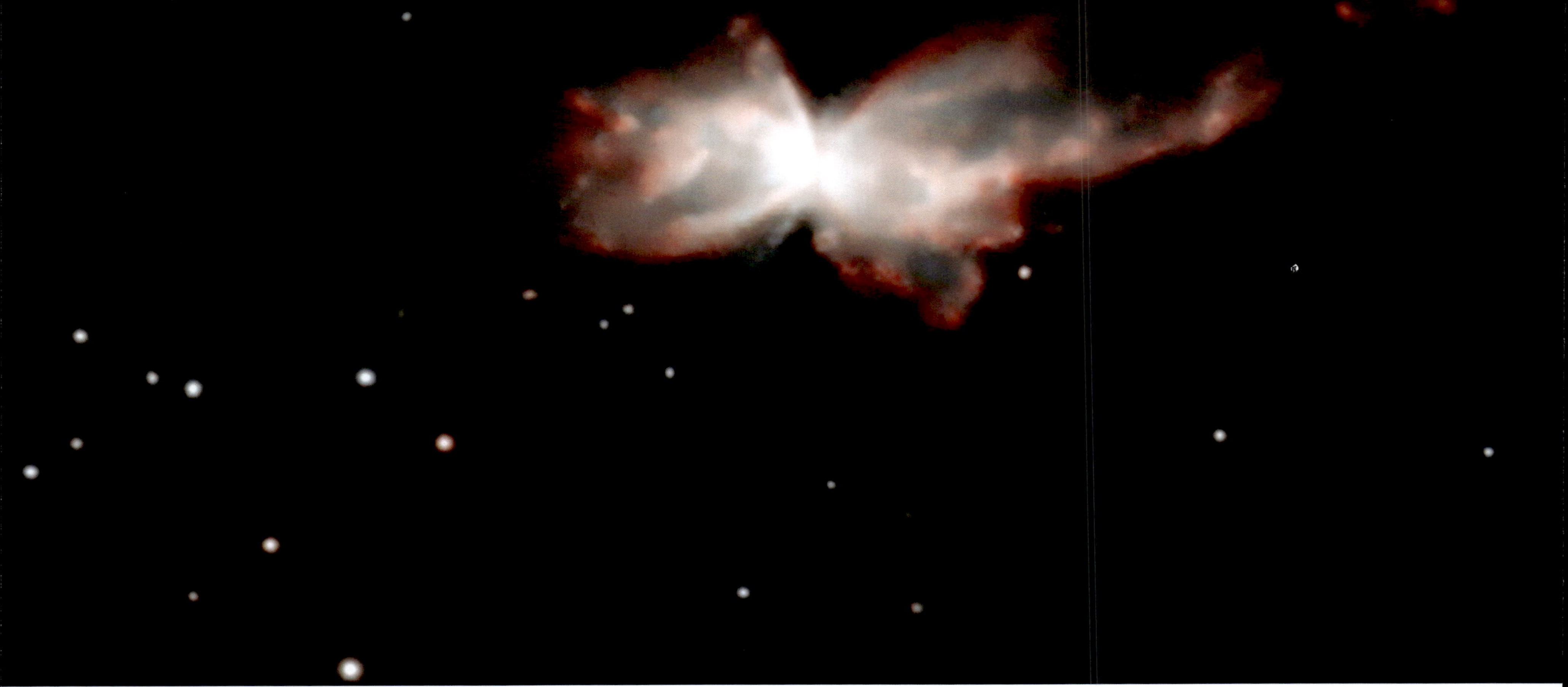

Bug / Butterfly Nebula (NGC 6302)
A bipolar planetary nebula in the constellation Scorpius
Approximate distance from Earth: 5,200 light-years

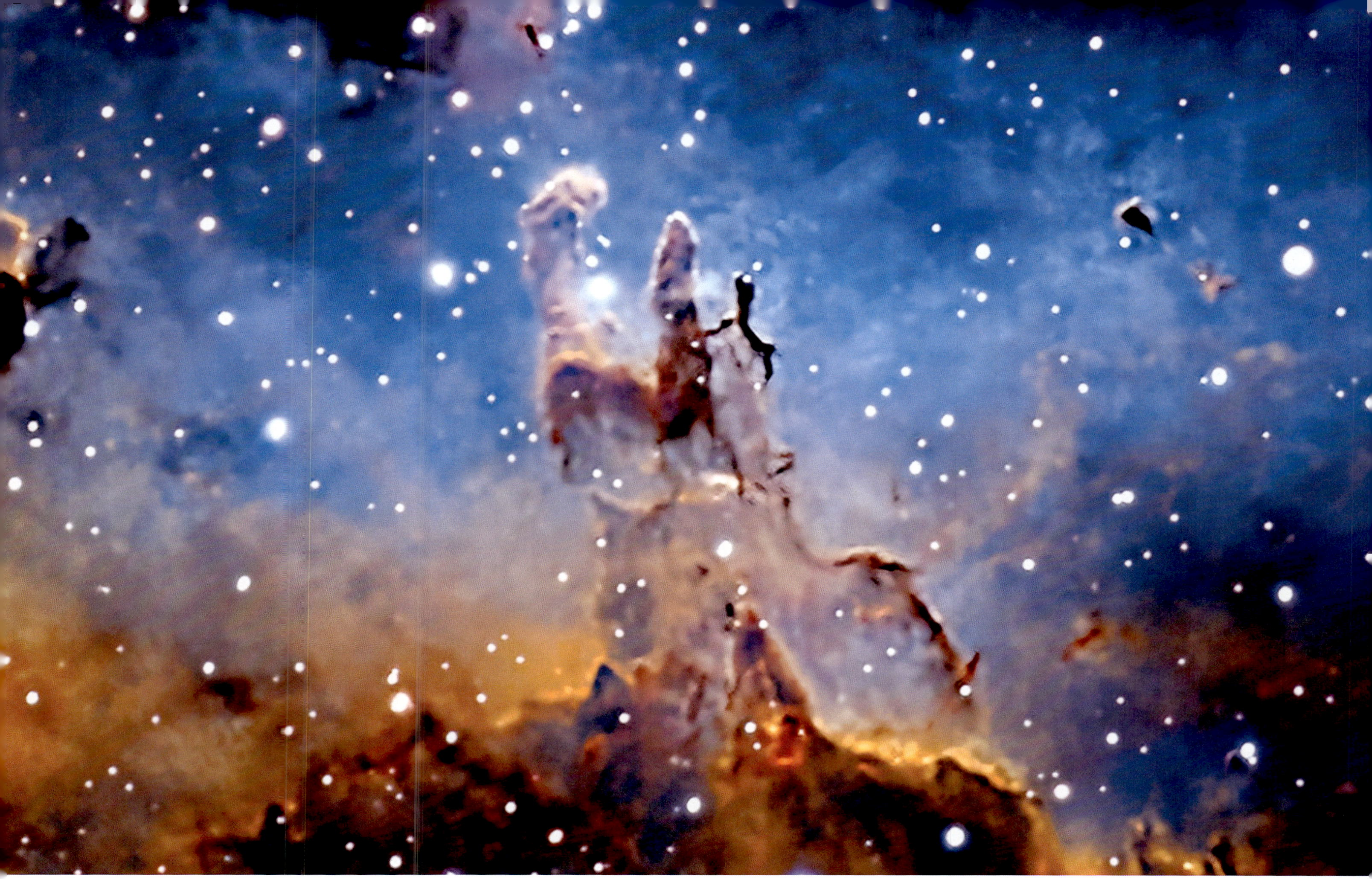

Messier 16 (The Eagle Nebula – "Pillars of Creation")
An emission nebula, the "Pillars of Creation" area is a stellar nursery / active star-forming region in the constellation Serpens
Approximate distance from Earth: 5,700 light-years

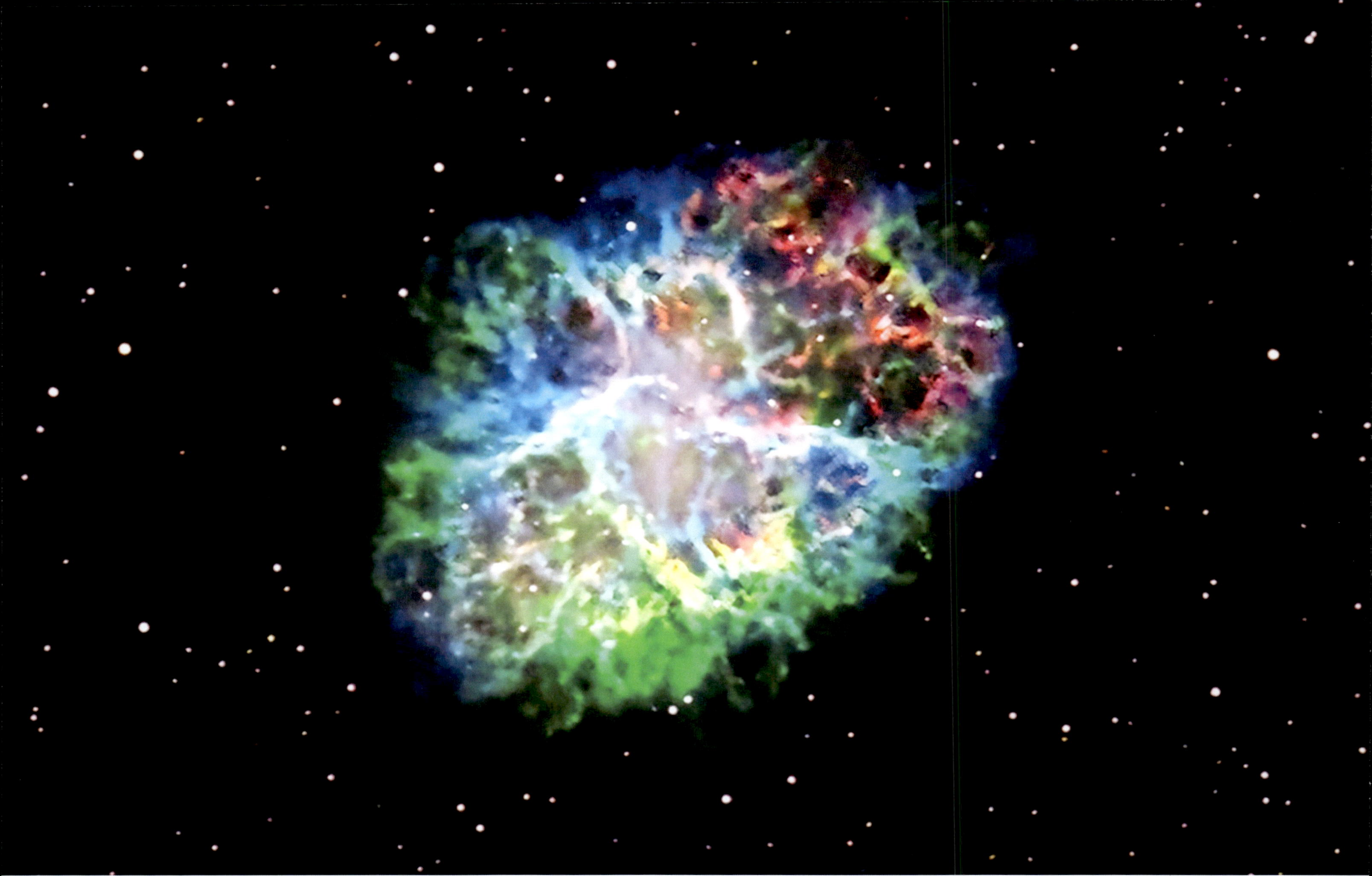

Messier 1 (The Crab Nebula)
A supernova remnant in the constellation Taurus
Approximate distance from Earth: 6,500 light-years

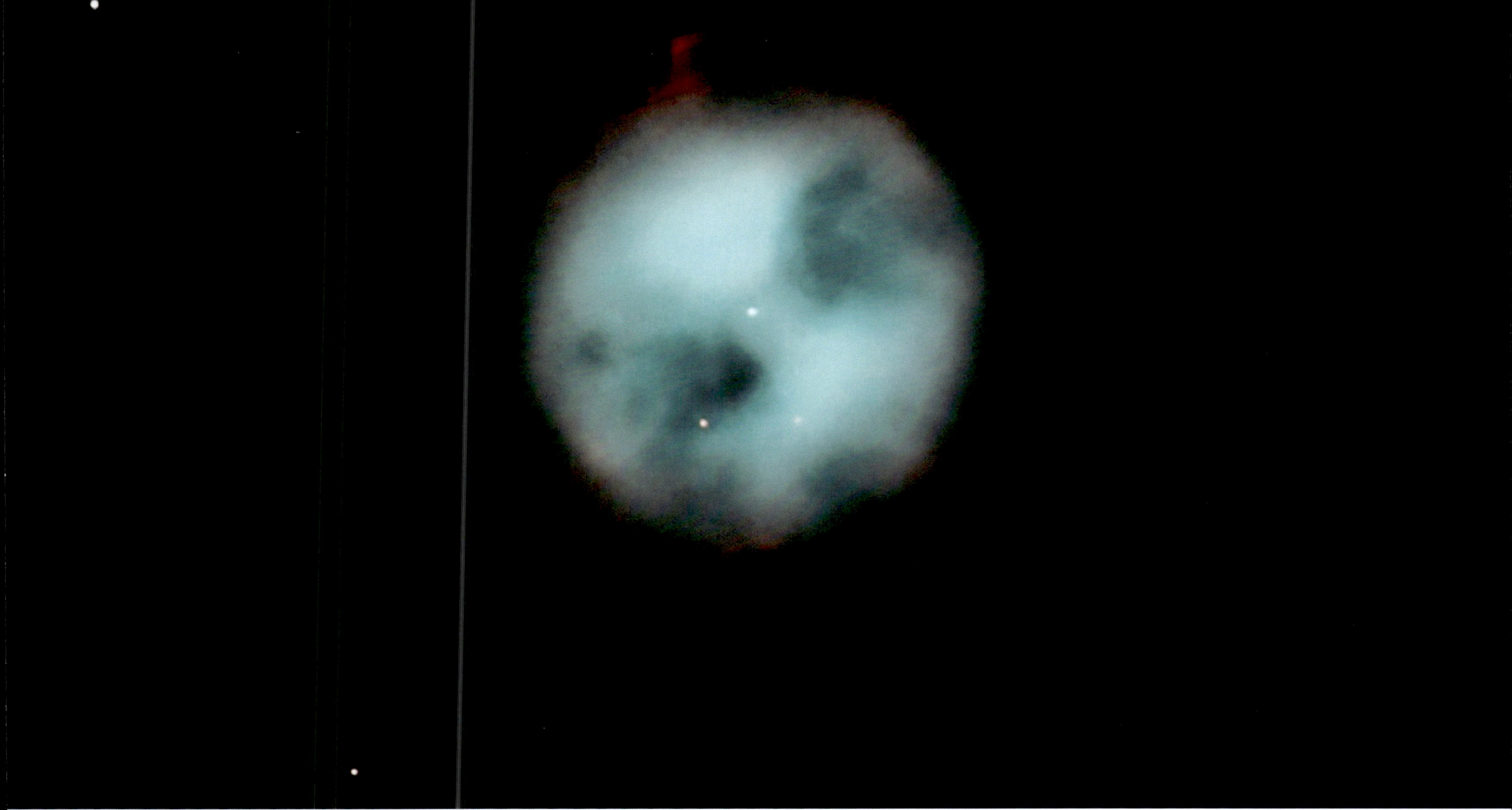

Messier 97 (The Owl Nebula)
A planetary nebula in the constellation Ursa Major
Approximate distance from Earth: 2,000 light-years

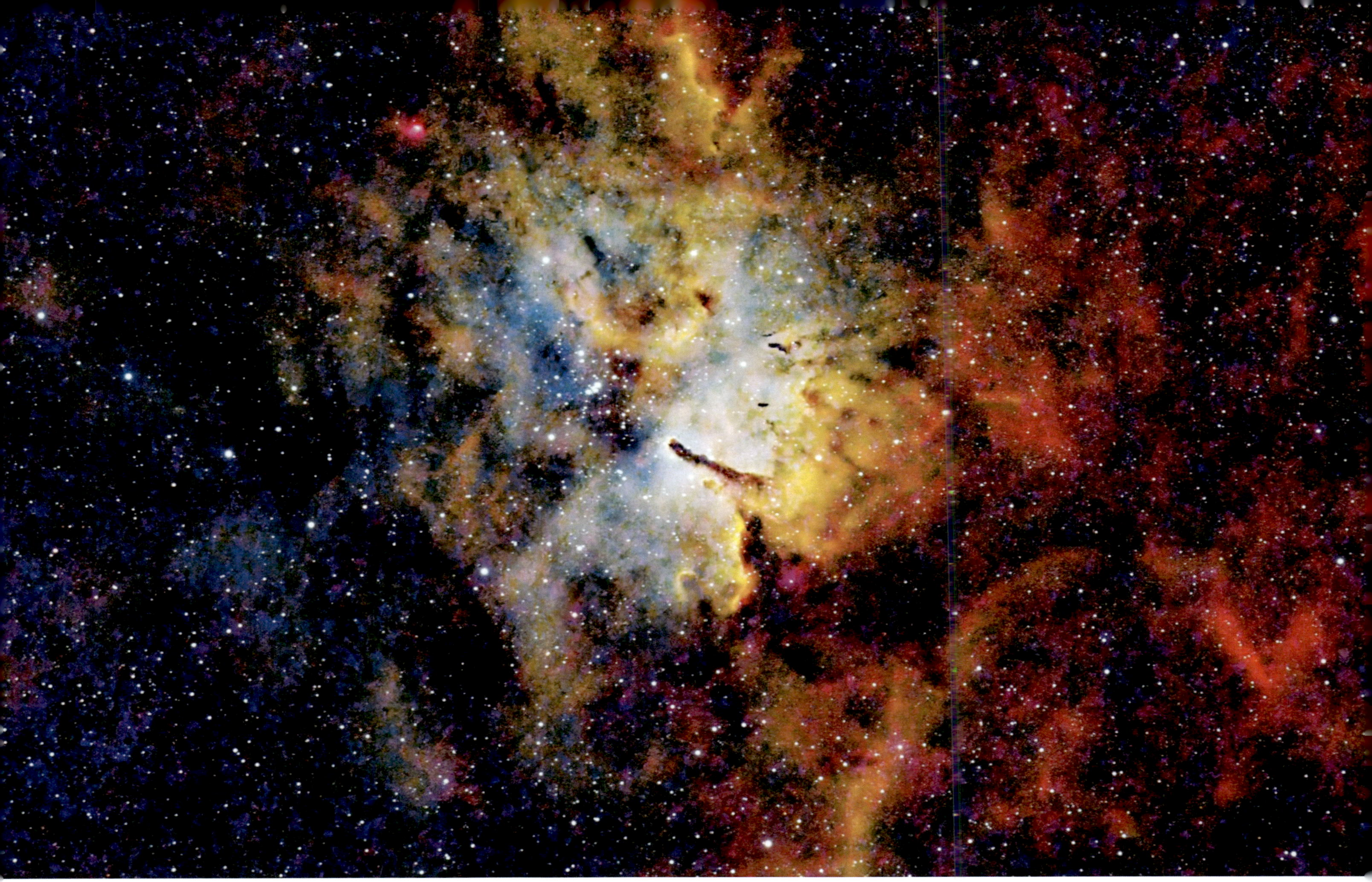

Sh2-86 (NGC 6820 / NGC 6823)
An emission nebula / star-forming region in the constellation Vulpecula
Approximate distance from Earth: 5,600 light-years

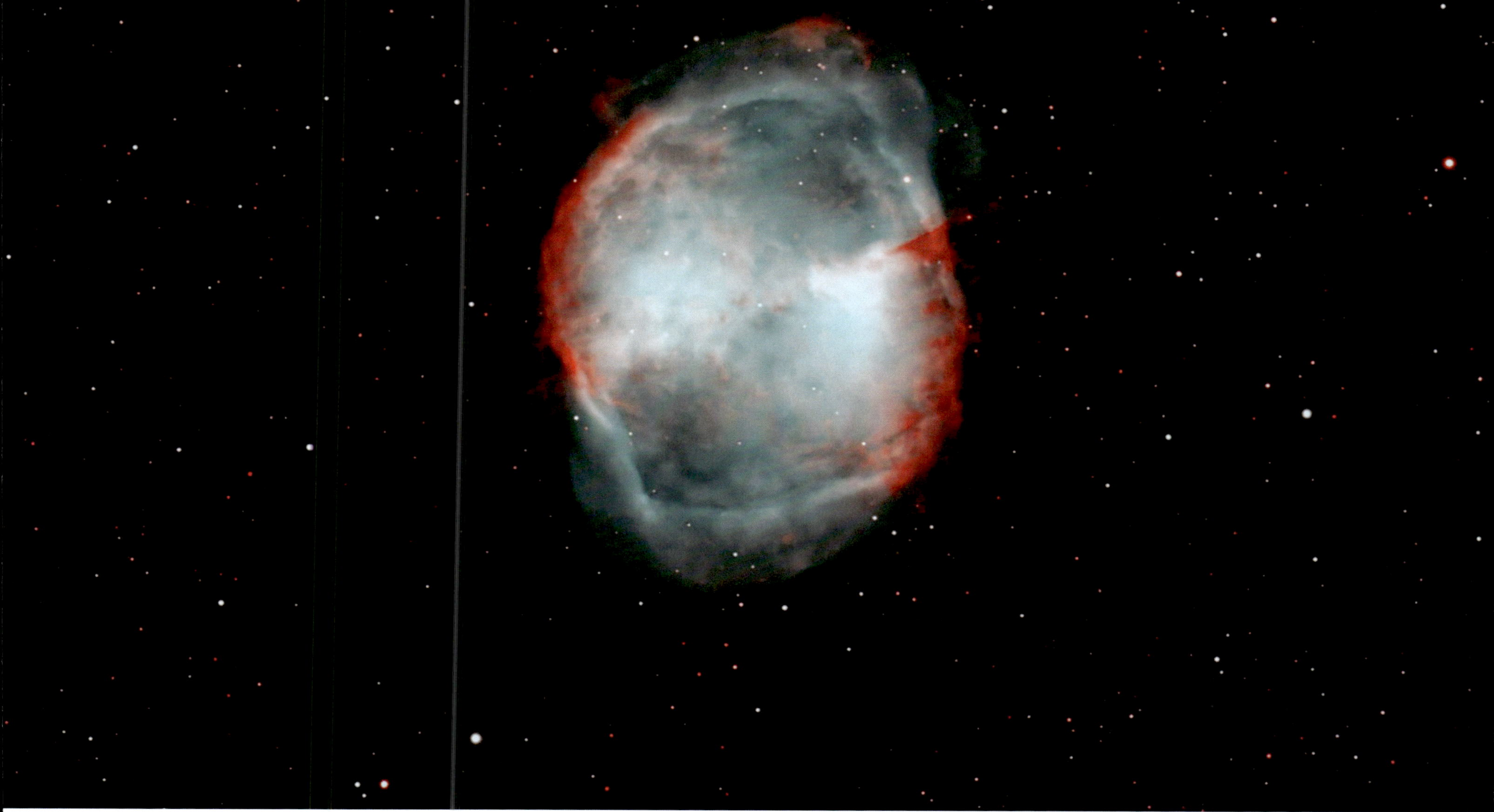

Messier 27 (The Dumbbell Nebula)
A planetary nebula in the constellation Vulpecula
Approximate distance from Earth: 1,360 light-years

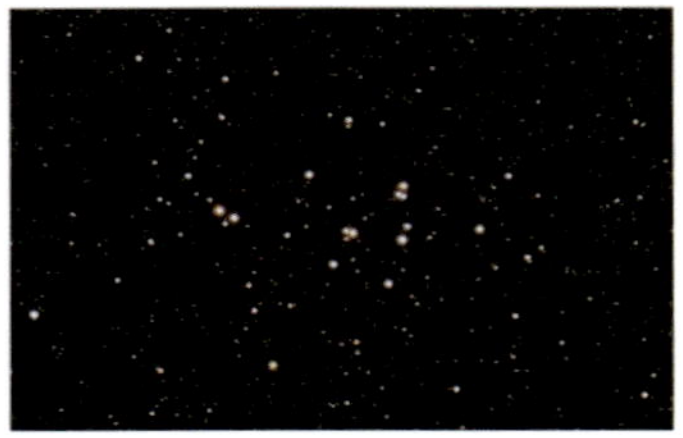

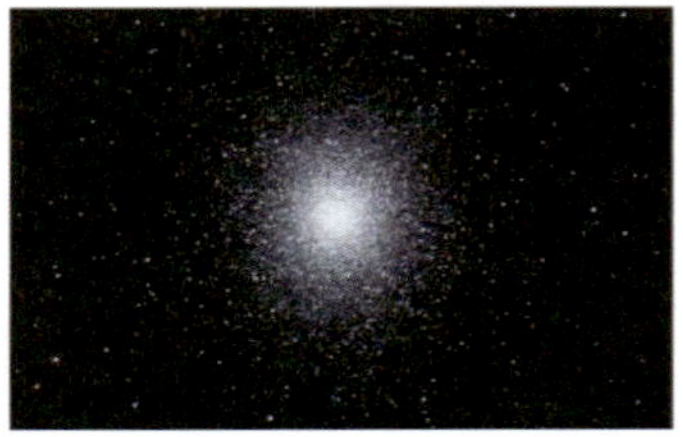

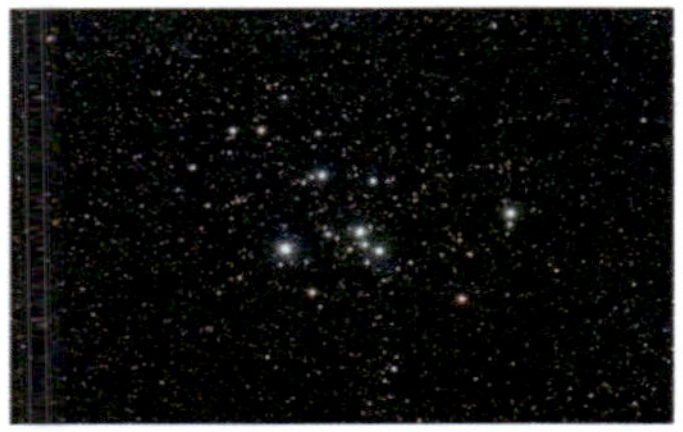

Star Cluster Images

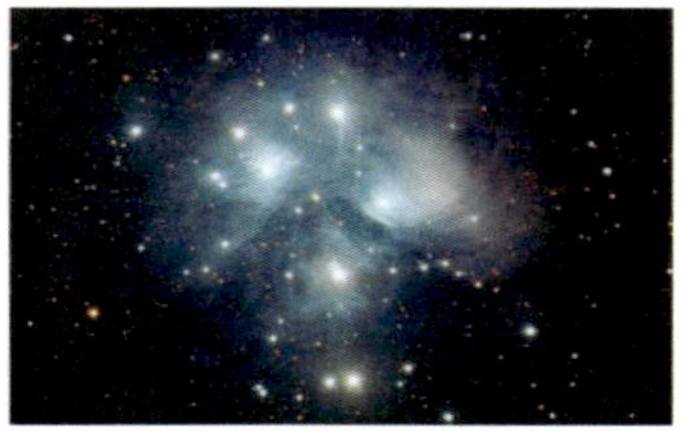

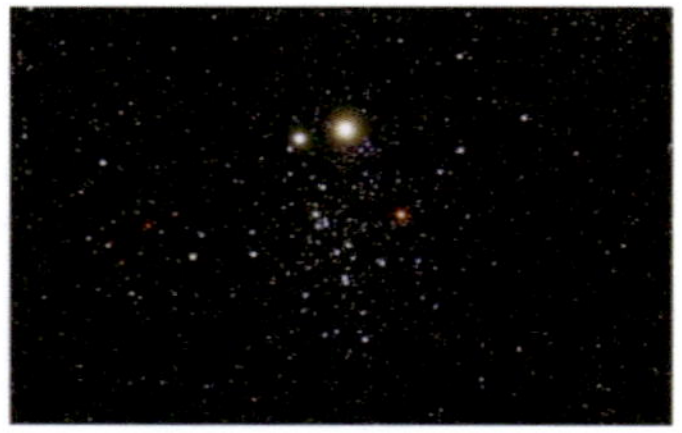

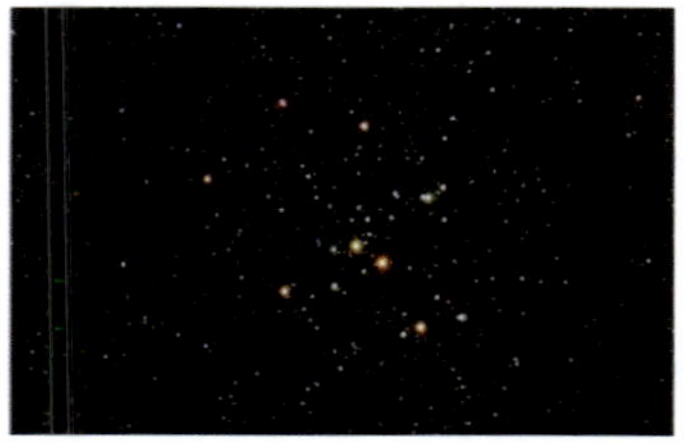

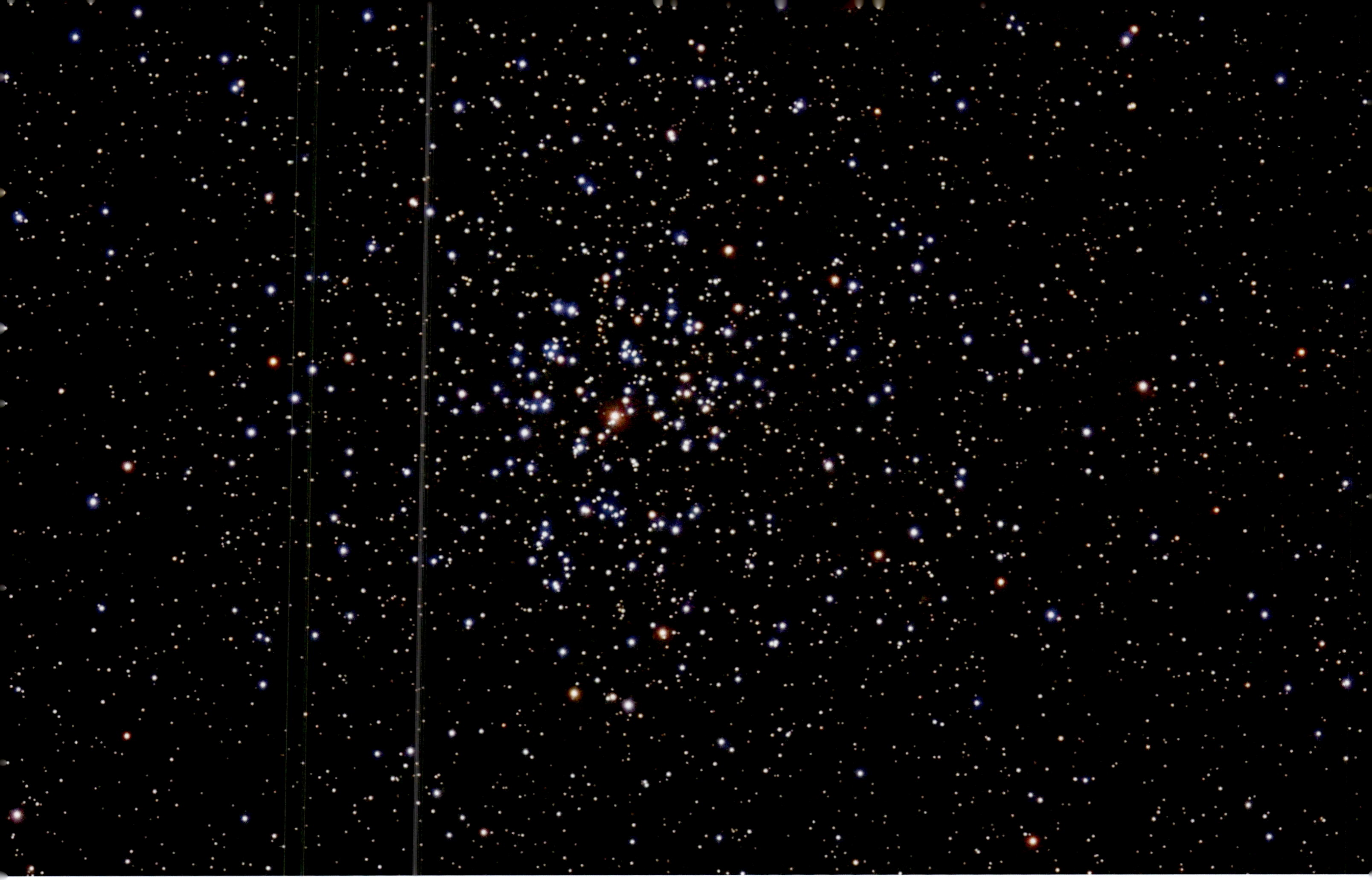

Messier 37
An open star cluster in the constellation Auriga
Approximate distance from Earth: 4,500 light-years

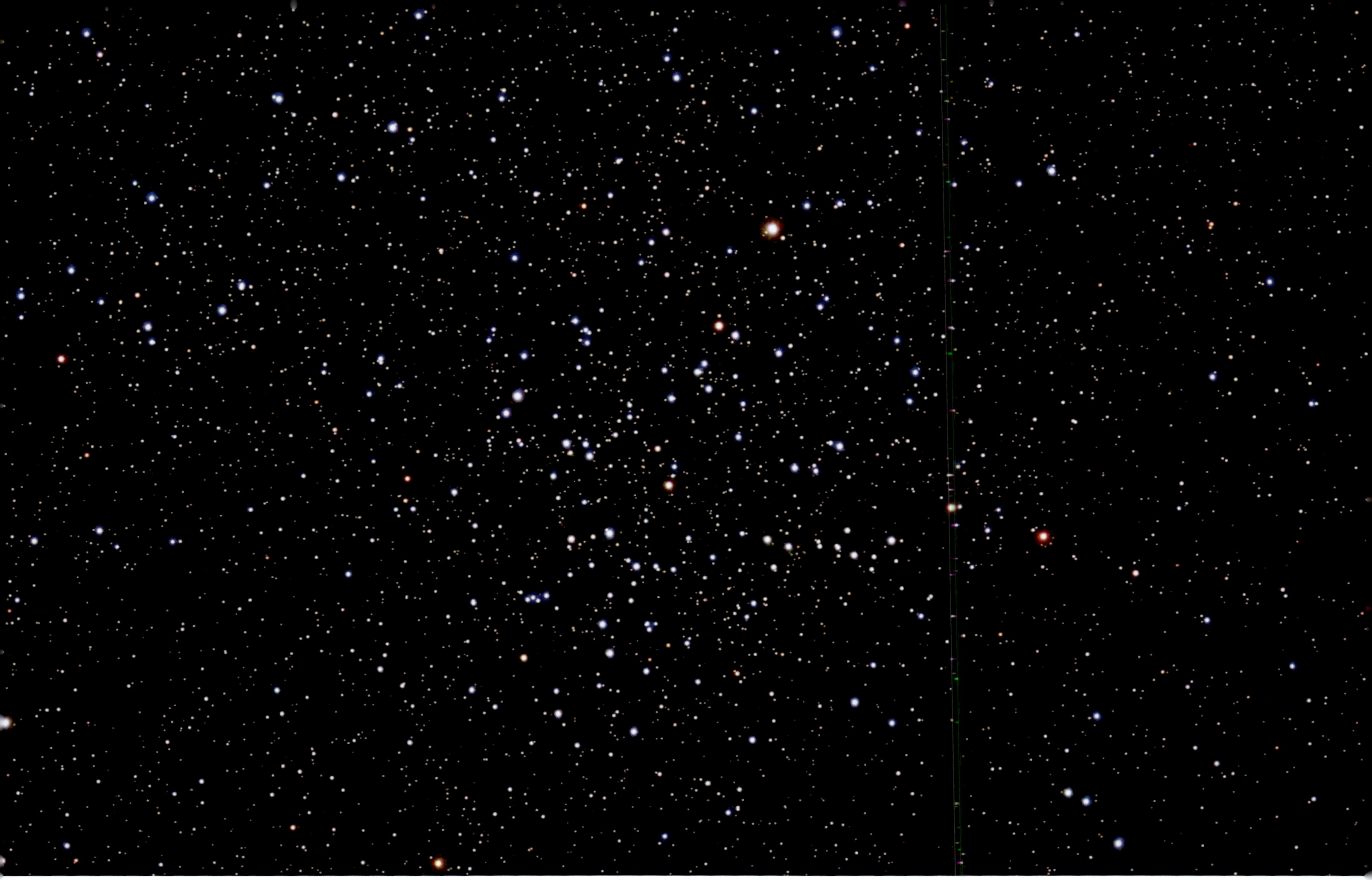

Messier 38
An open star cluster in the constellation Auriga
Approximate distance from Earth: 3,500 light-years

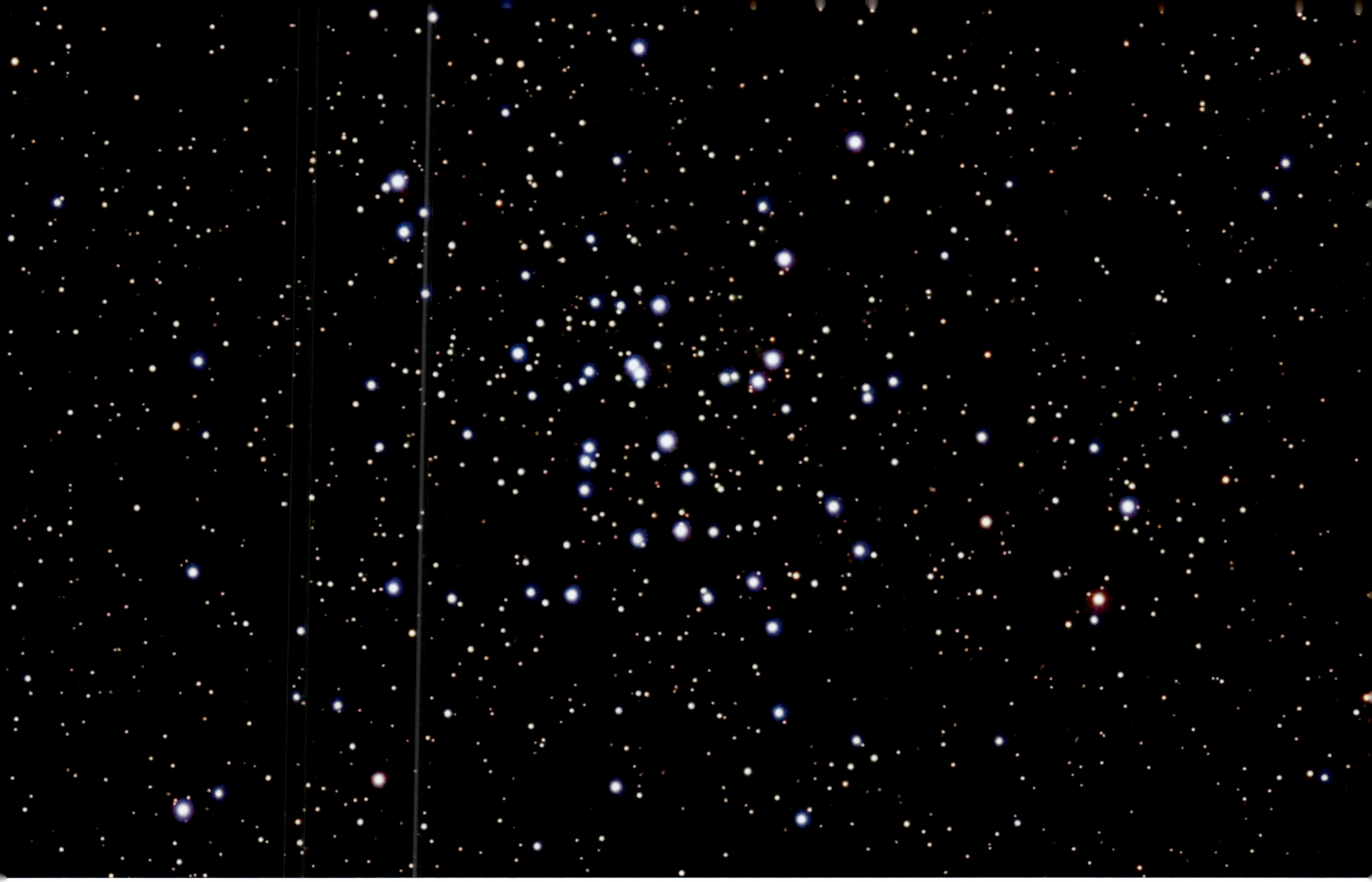

Messier 36
An open star cluster in the constellation Auriga
Approximate distance from Earth: 4,100 light-years

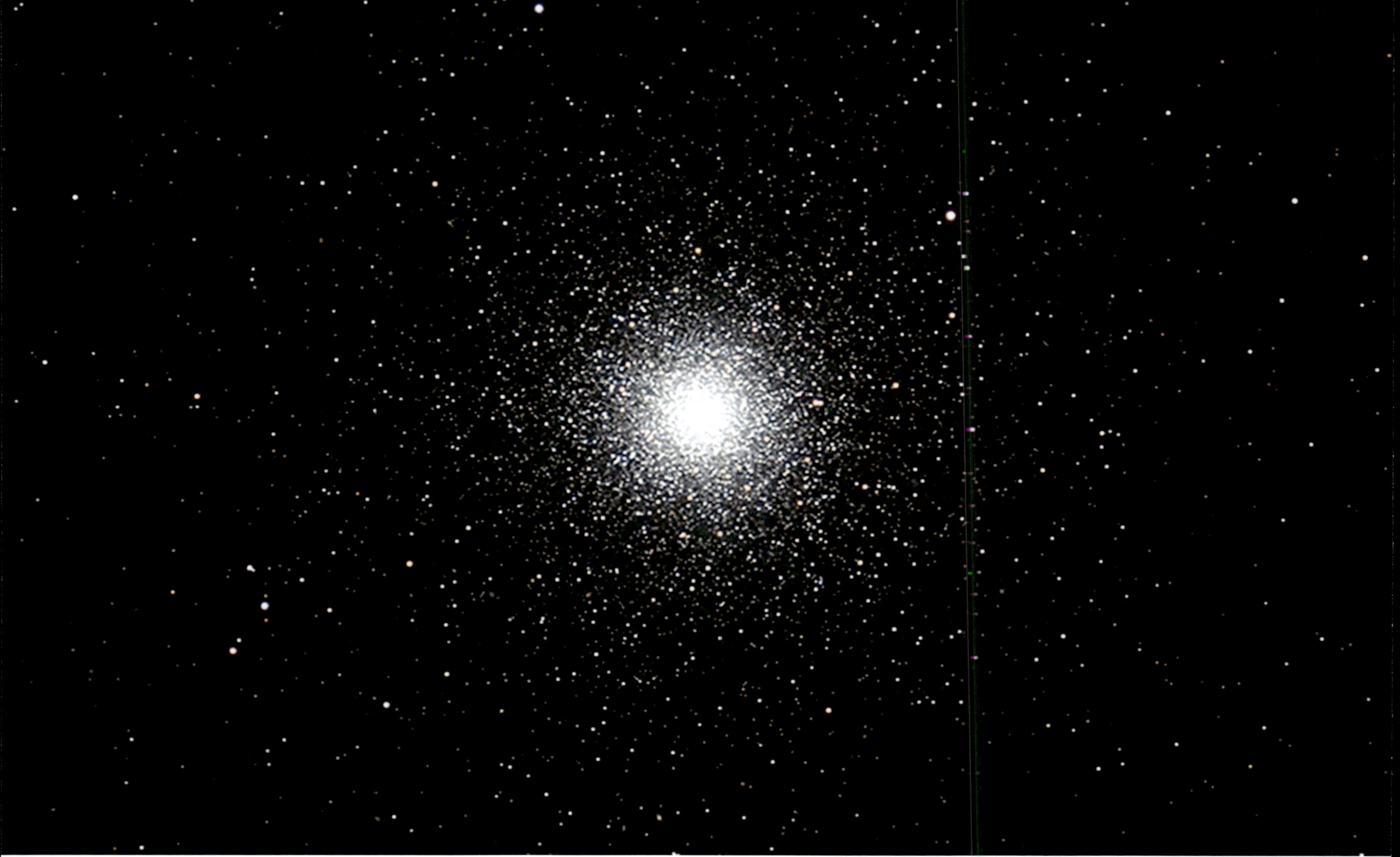

Messier 3
A globular star cluster in the constellation Bootes
Approximate distance from Earth: 34,000 light-years

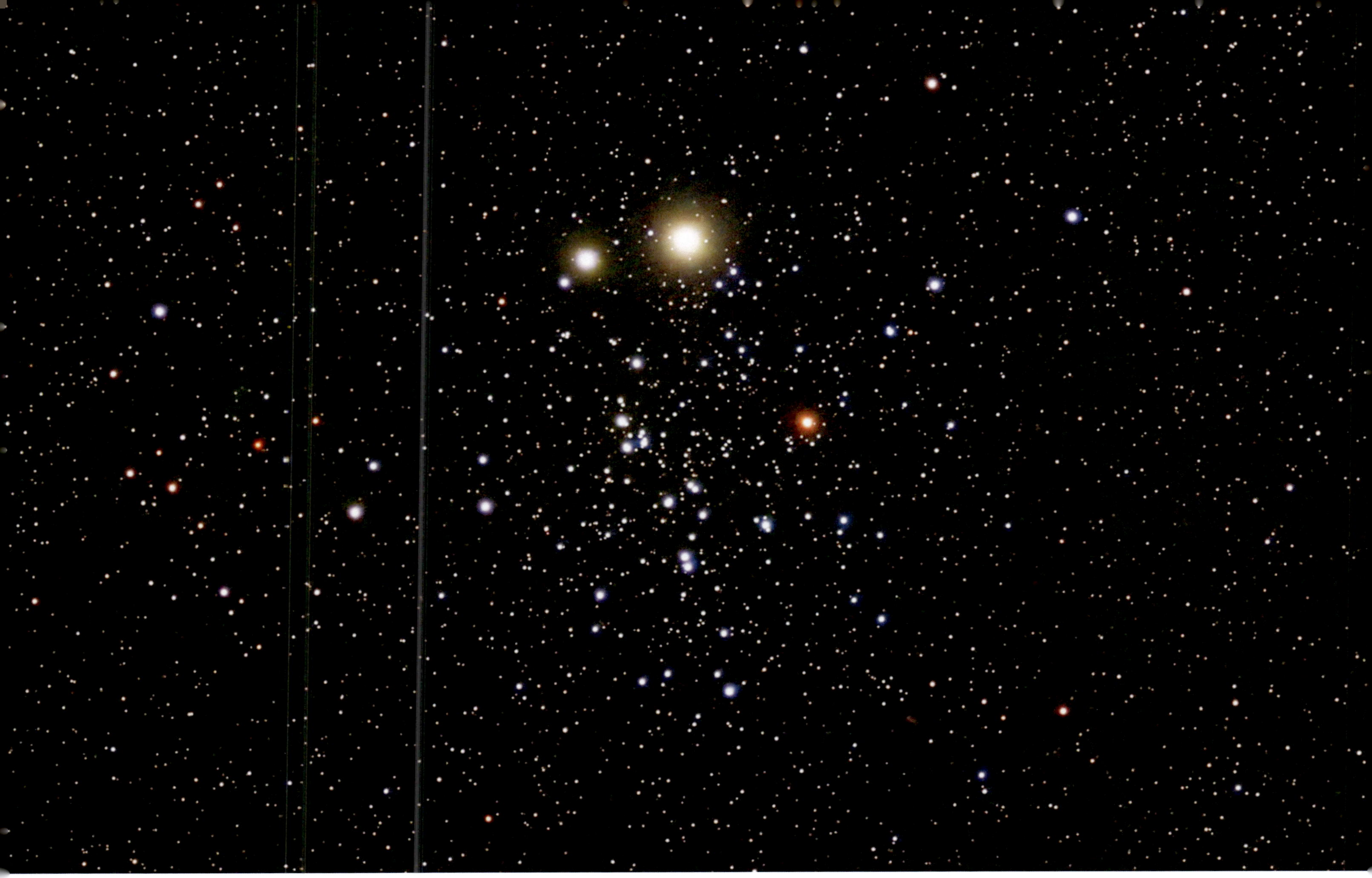

The Owl/ET Cluster (NGC 457)
An open star cluster in the constellation Cassiopeia
Approximate distance from Earth: 7,900 light-years

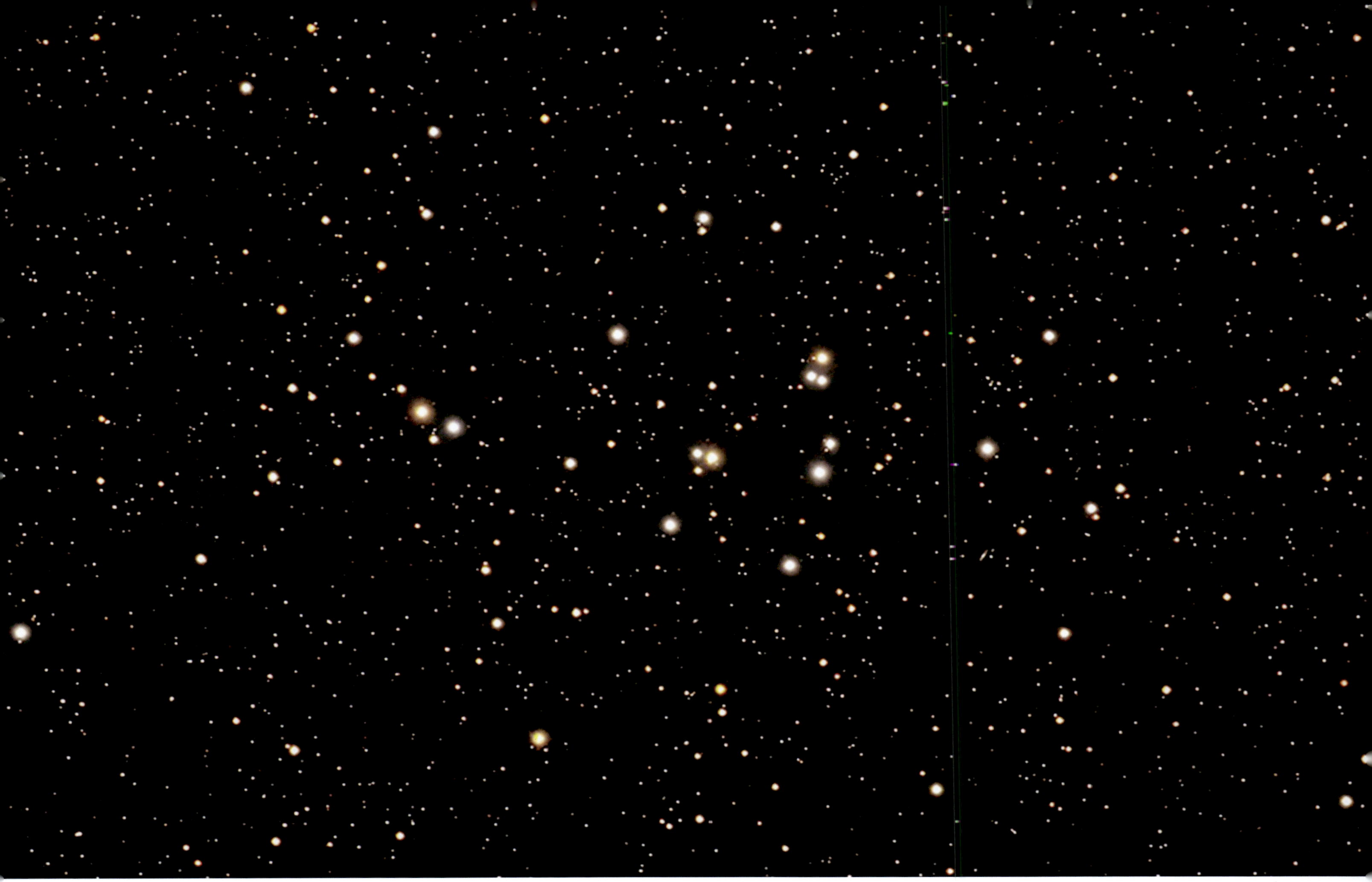

Messier 44 (Beehive Cluster)
An open star cluster in the constellation Cancer
Approximate distance from Earth: 610 light-years

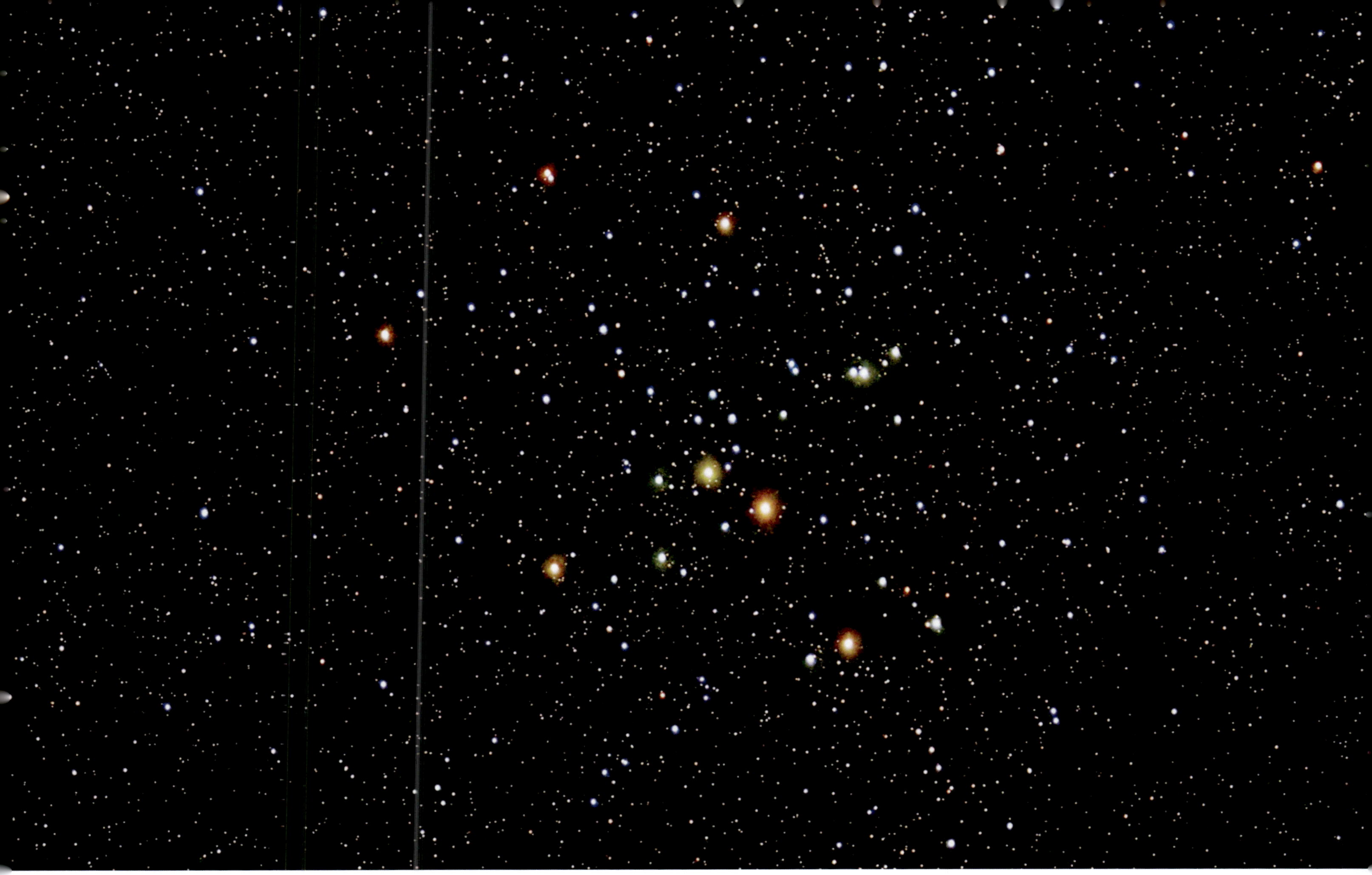

Messier 41 (The Little Beehive Cluster)
An open star cluster in the constellation Canis Major
Approximate distance from Earth: 2,300 light-years

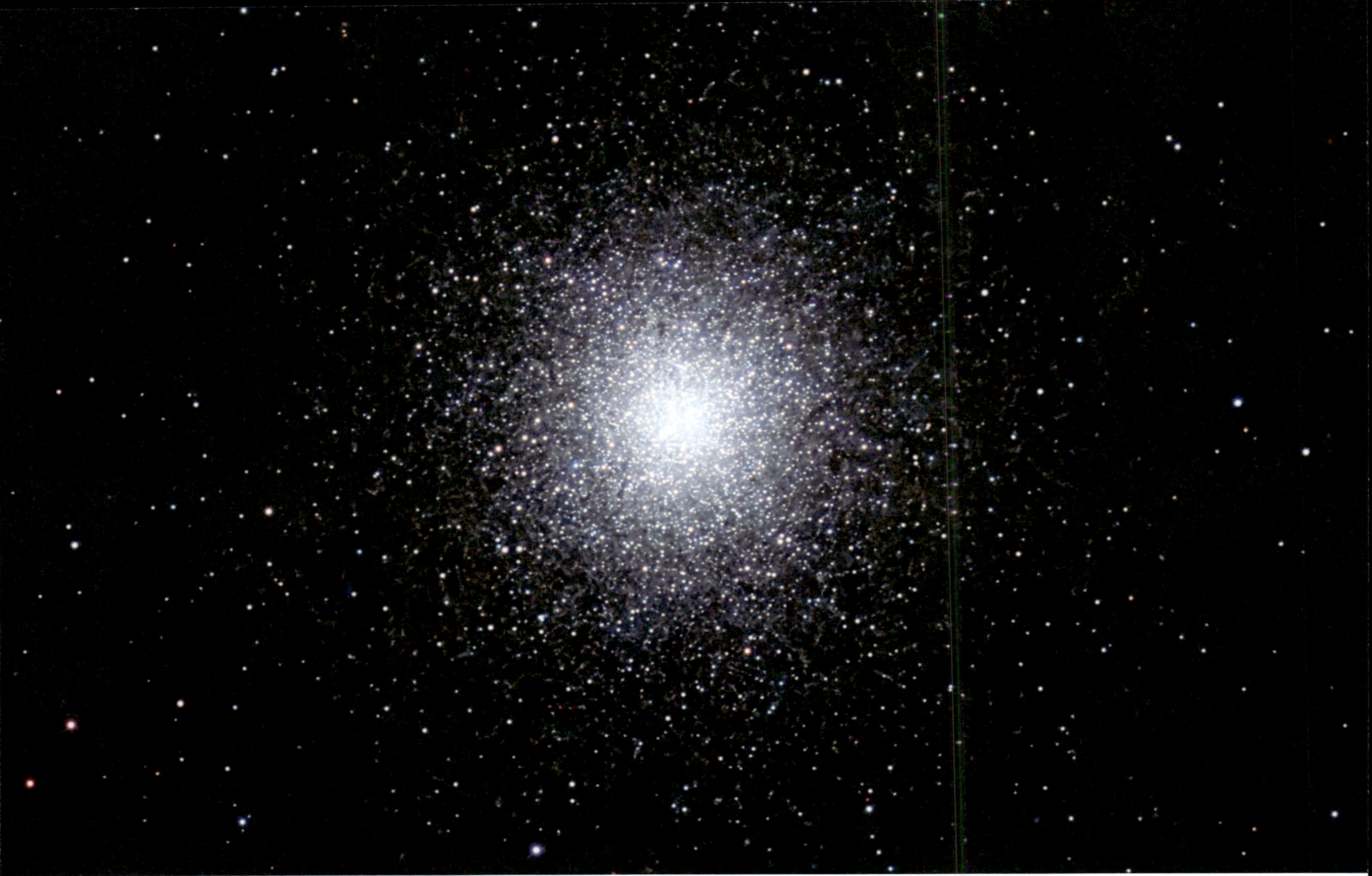

Messier 13 (The Great Hercules Cluster)
A globular star cluster in the constellation Hercules
Approximate distance from Earth: 25,000 light-years

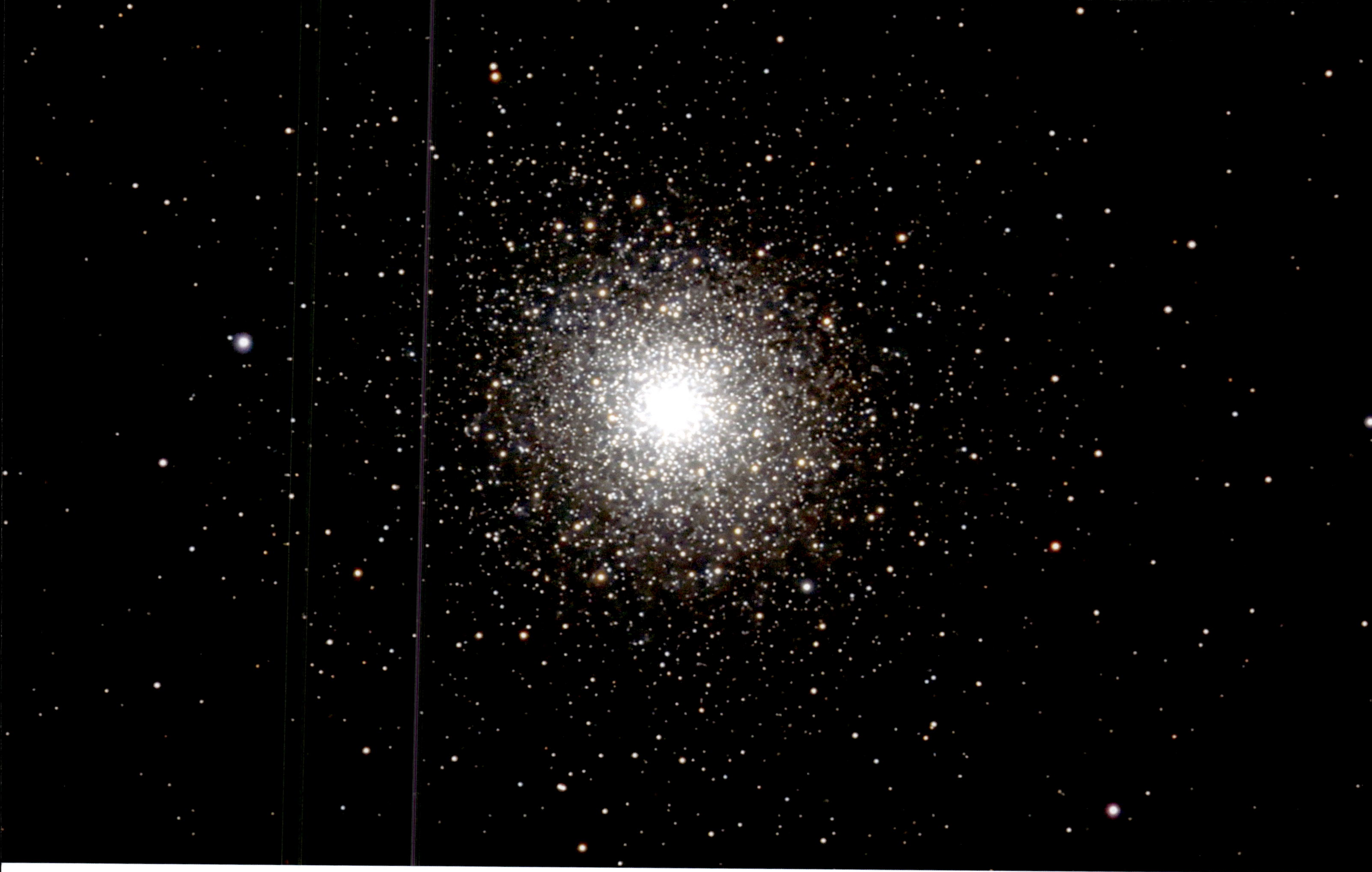

Messier 92
A globular star cluster in the constellation Hercules
Approximate distance from Earth: 27,000 light-years

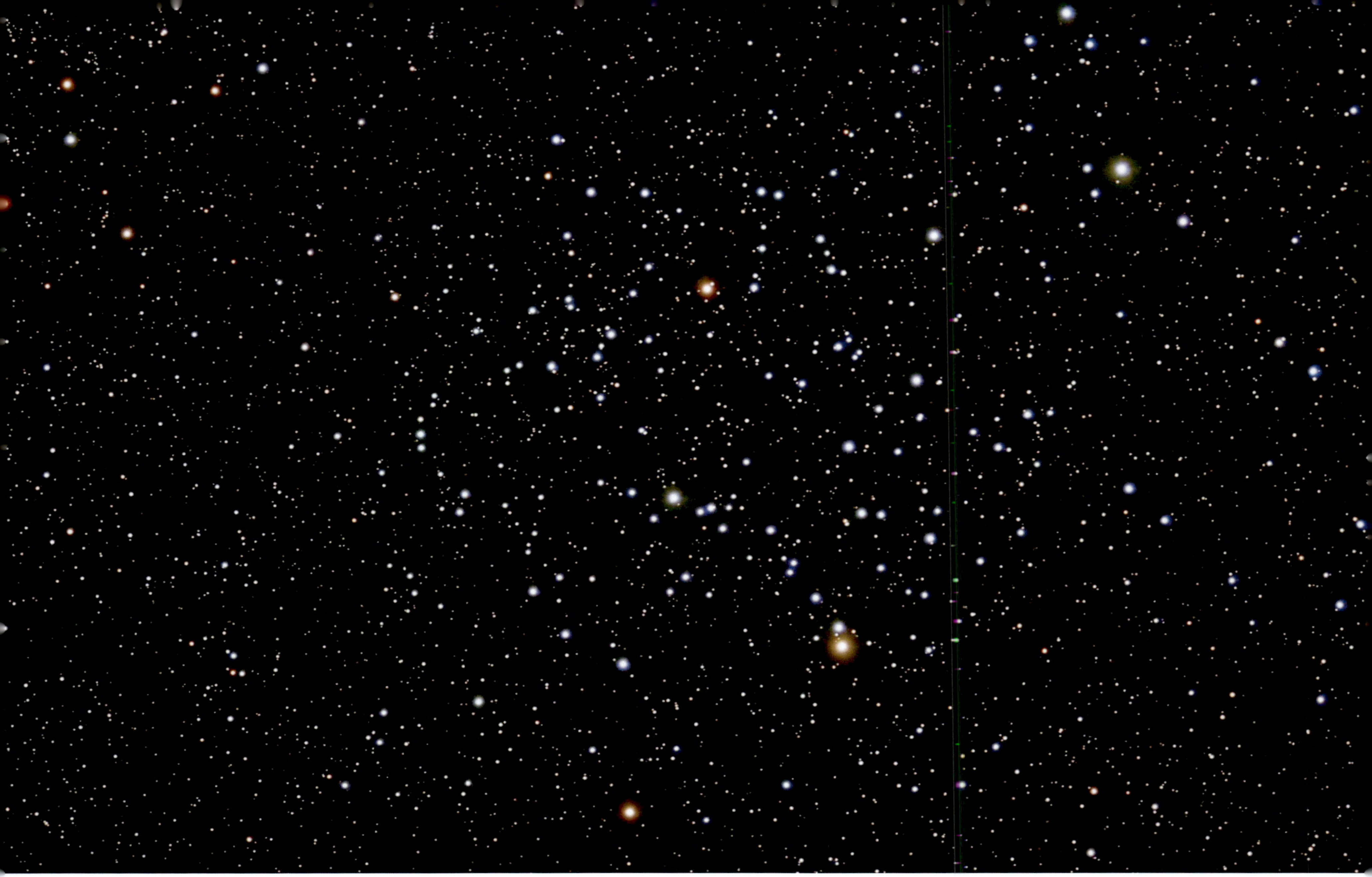

Messier 35
An open star cluster in the constellation Gemini
Approximate distance from Earth: 2,800 light-years

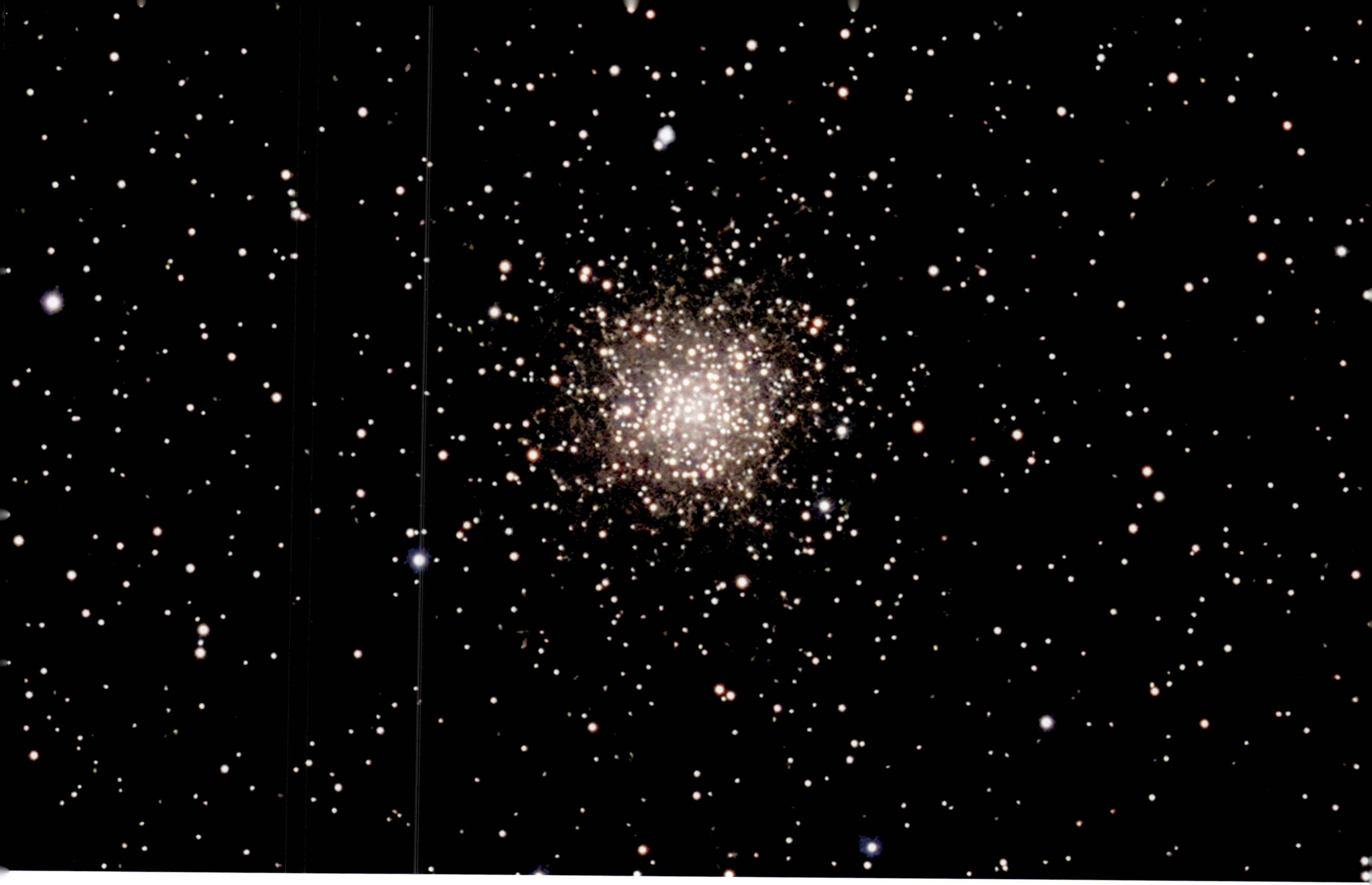

Messier 56
A globular star cluster in the constellation Lyra
Approximate distance from Earth: 33,000 light-years

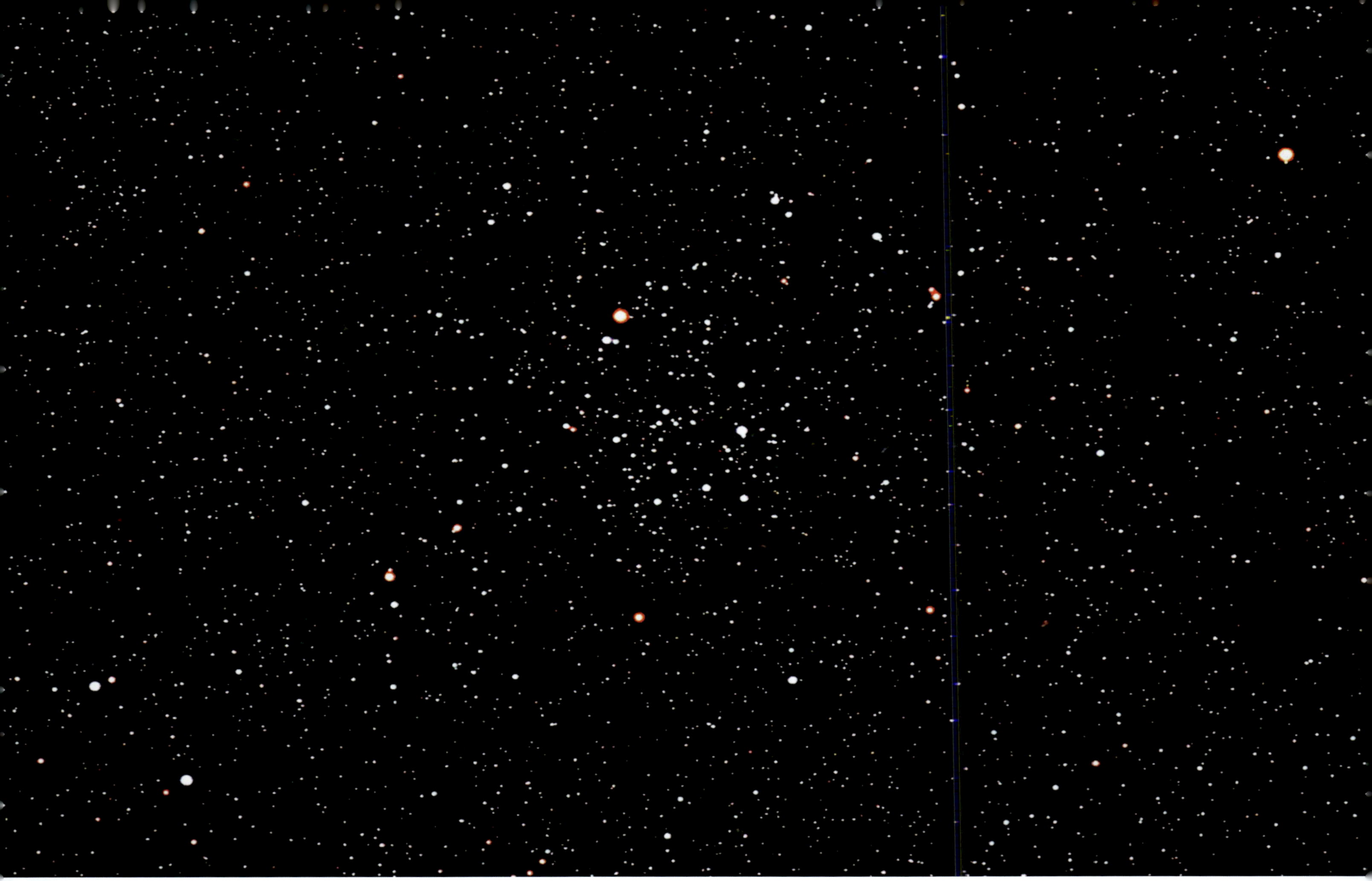

Messier 50
An open star cluster in the constellation Monoceros
Approximate distance from Earth: 3,200 light-years

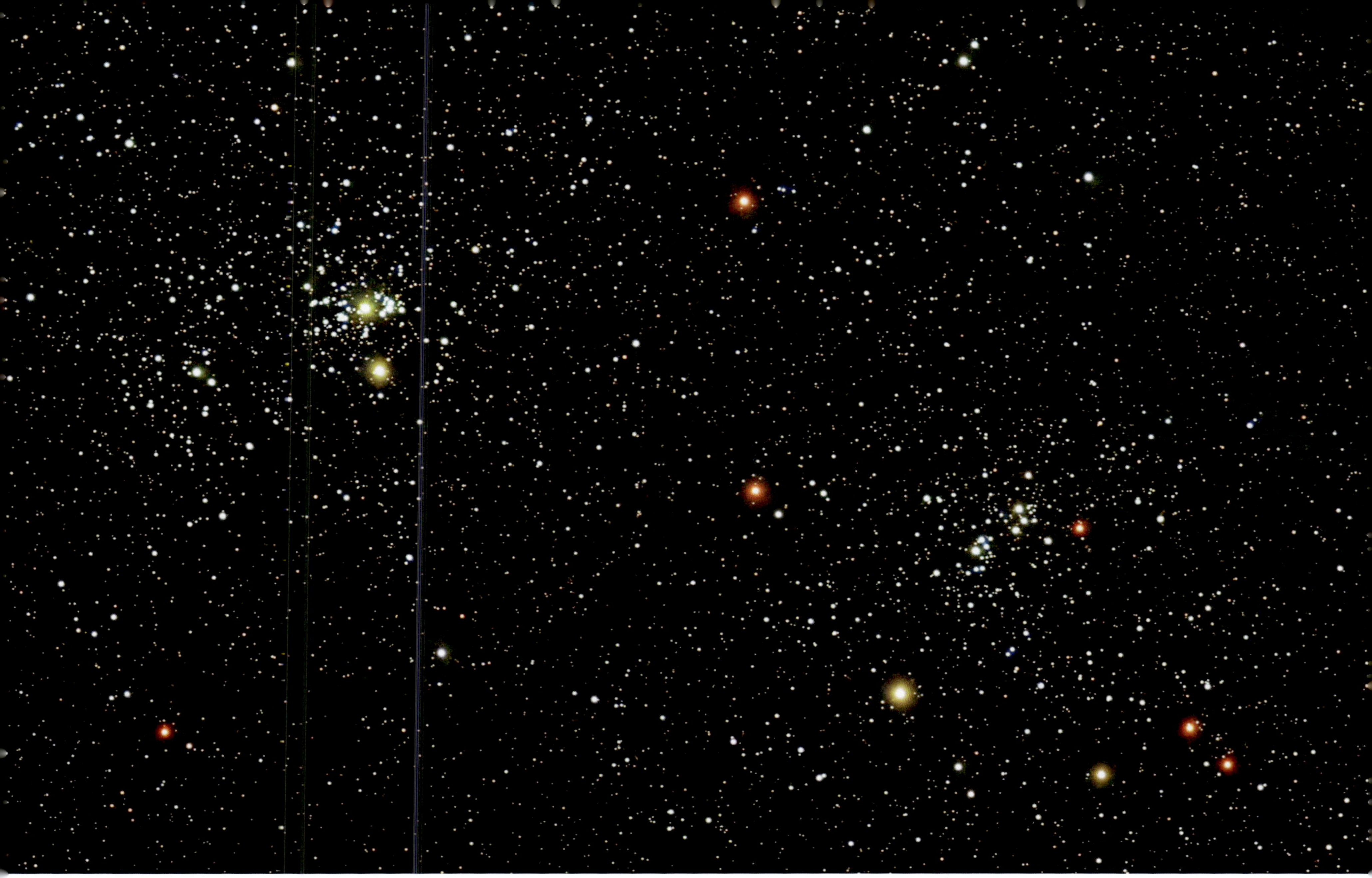

The Double Cluster (NGC 869 & NGC 884)
A pair of open star clusters in the constellation Perseus
Approximate distance from Earth: 7,500 light-years

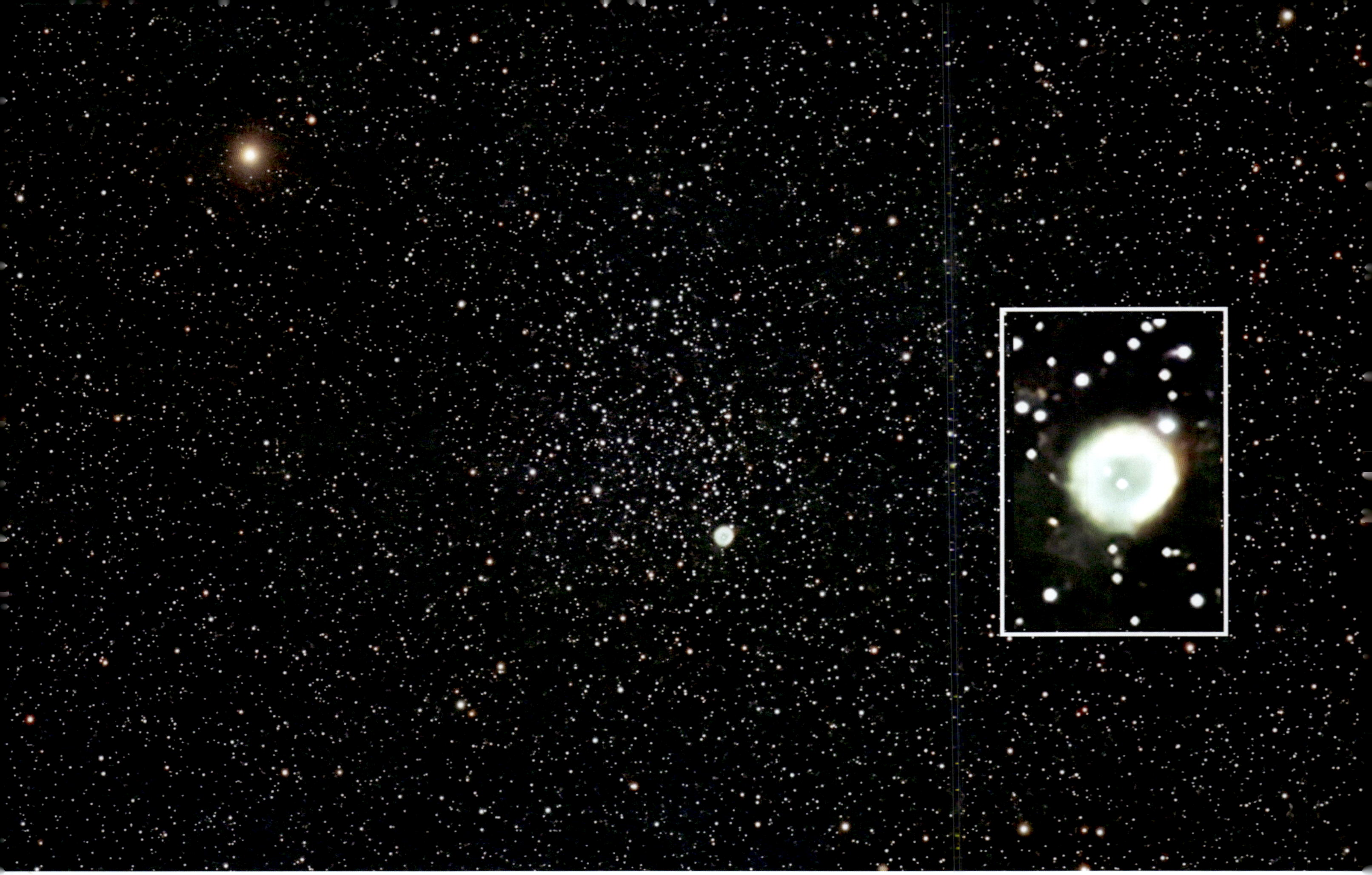

Messier 46 (with planetary nebula NGC 2438 – a foreground object)
An open star cluster and a planetary nebula in the constellation Puppis
Approximate distance from Earth: 5,400 light-years

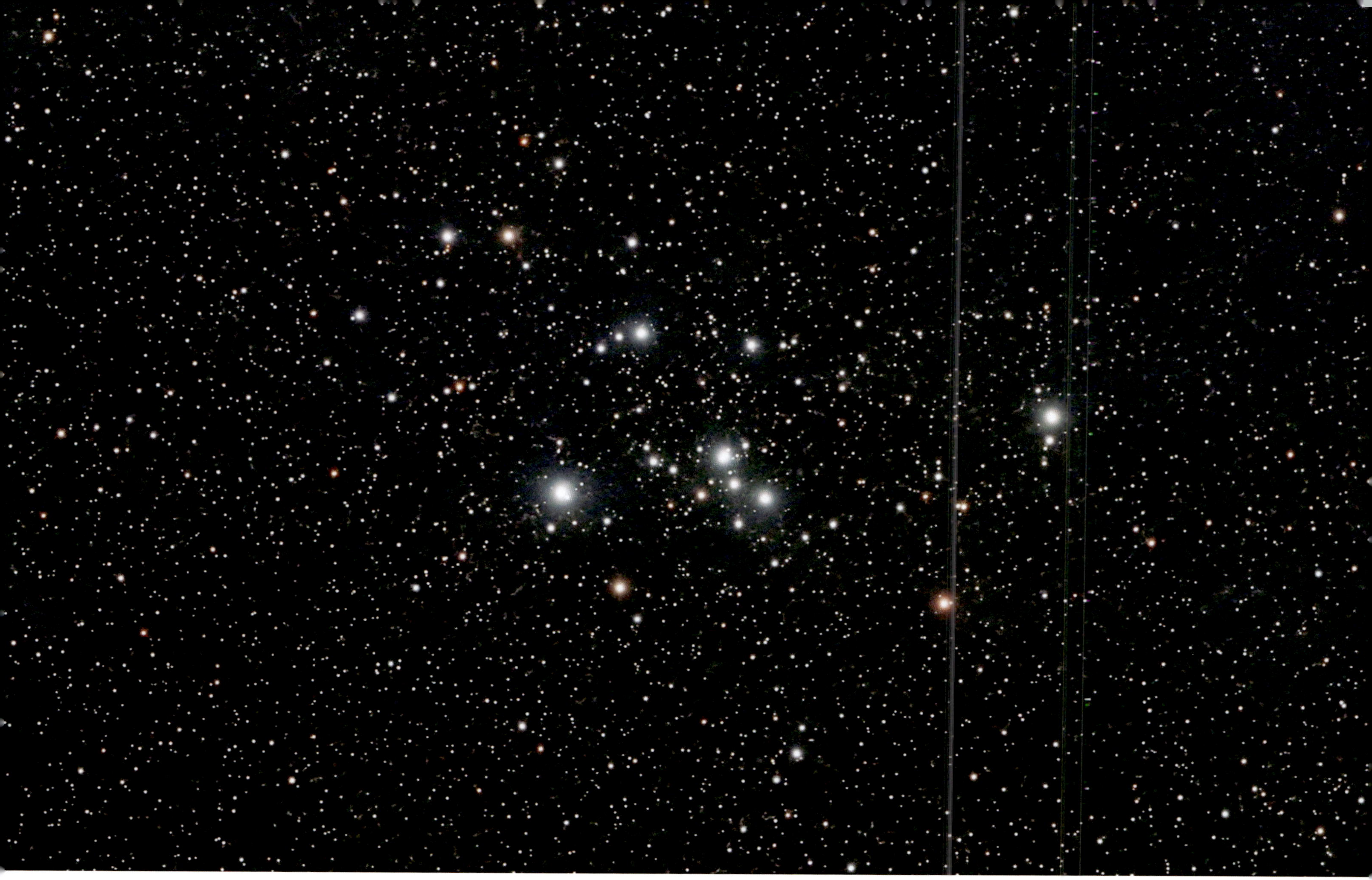

Messier 47
An open star cluster in the constellation Puppis
Approximate distance from Earth: 1,600 light-years

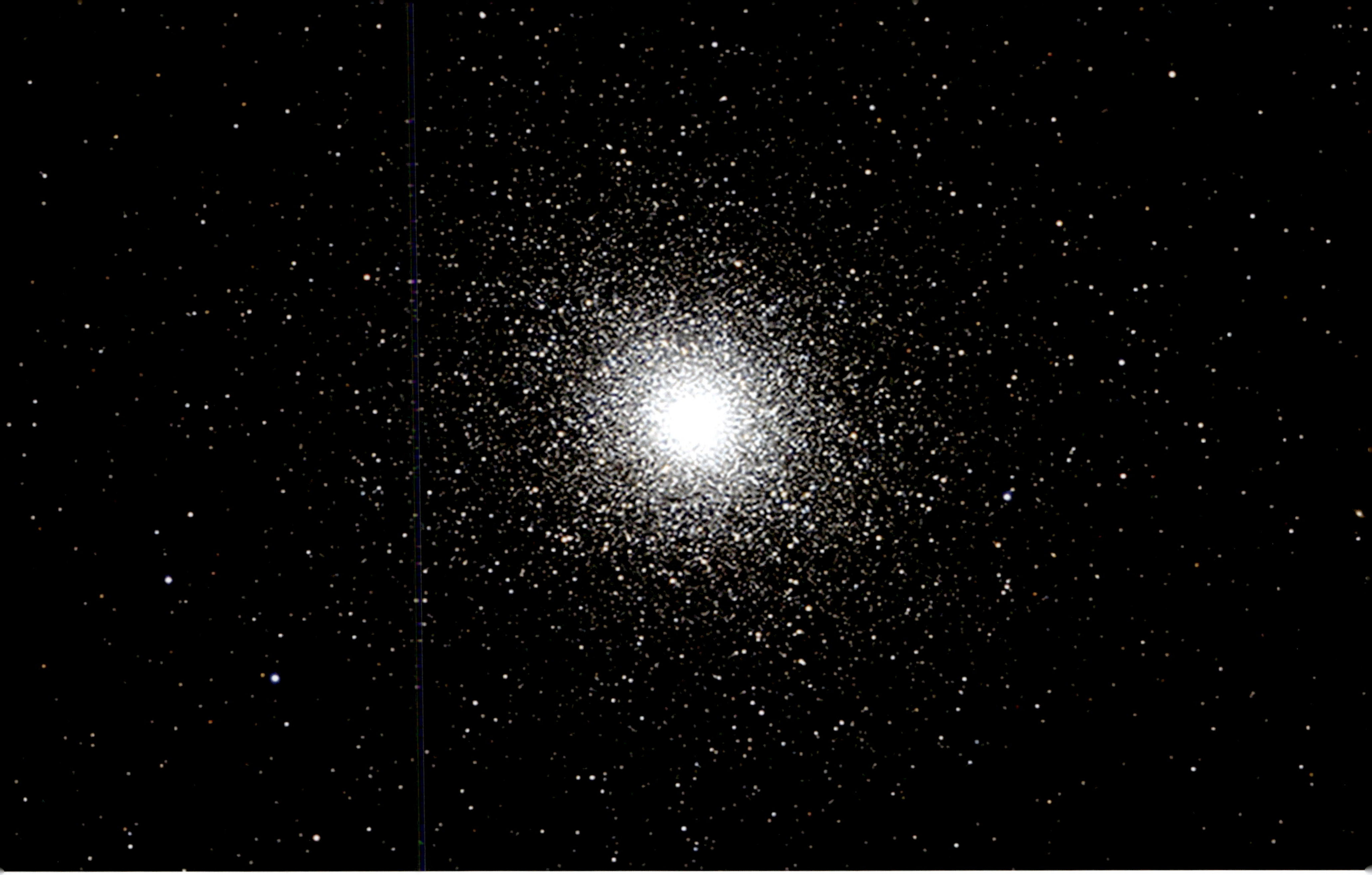

Messier 5
A globular star cluster in the constellation Serpens
Approximate distance from Earth: 24,500 light-years

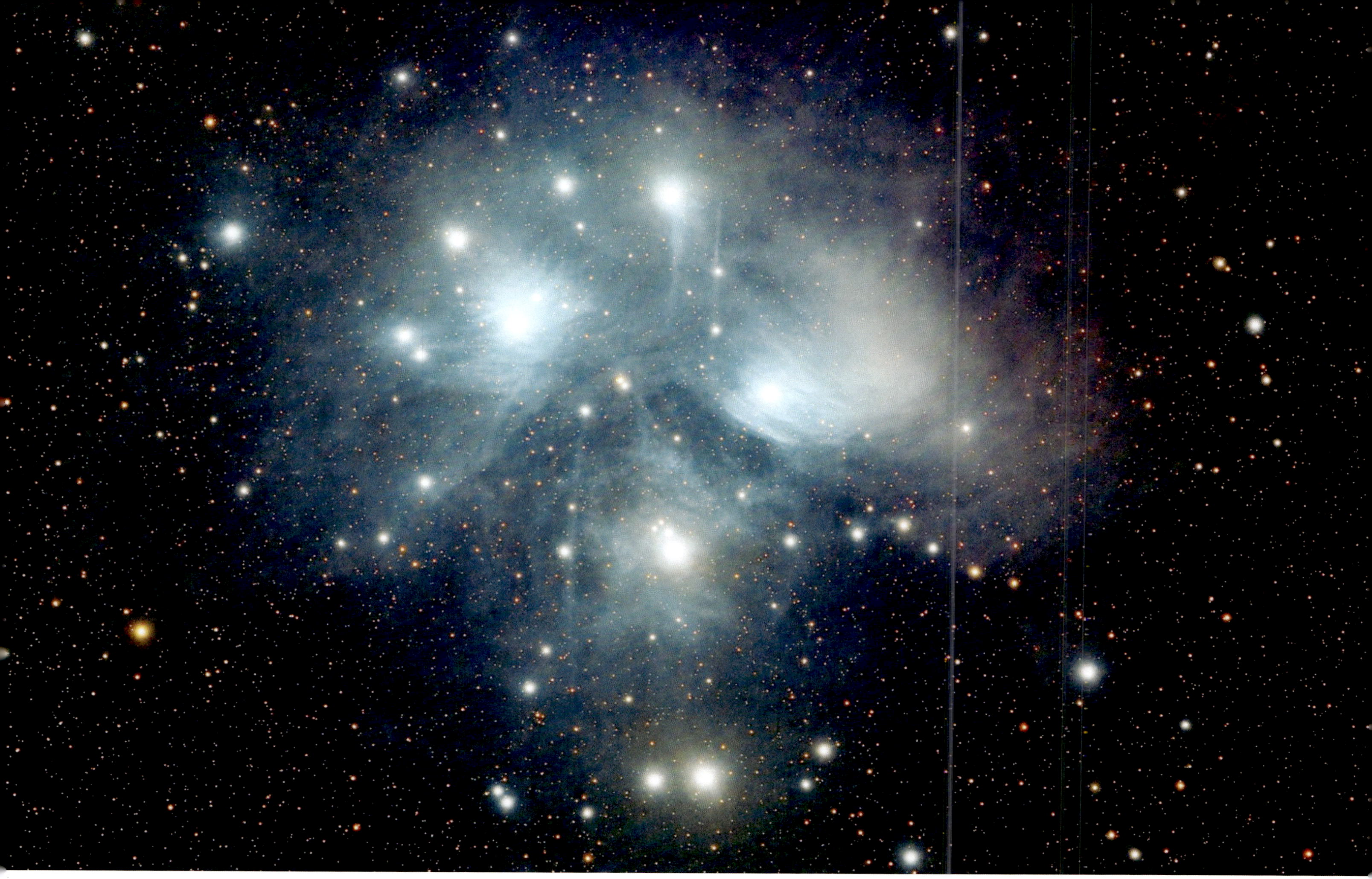

Messier 45 (The Pleiades – also known as the Seven Sisters)
An open star cluster and reflection nebula in the constellation Taurus
Approximate distance from Earth: 445 light-years

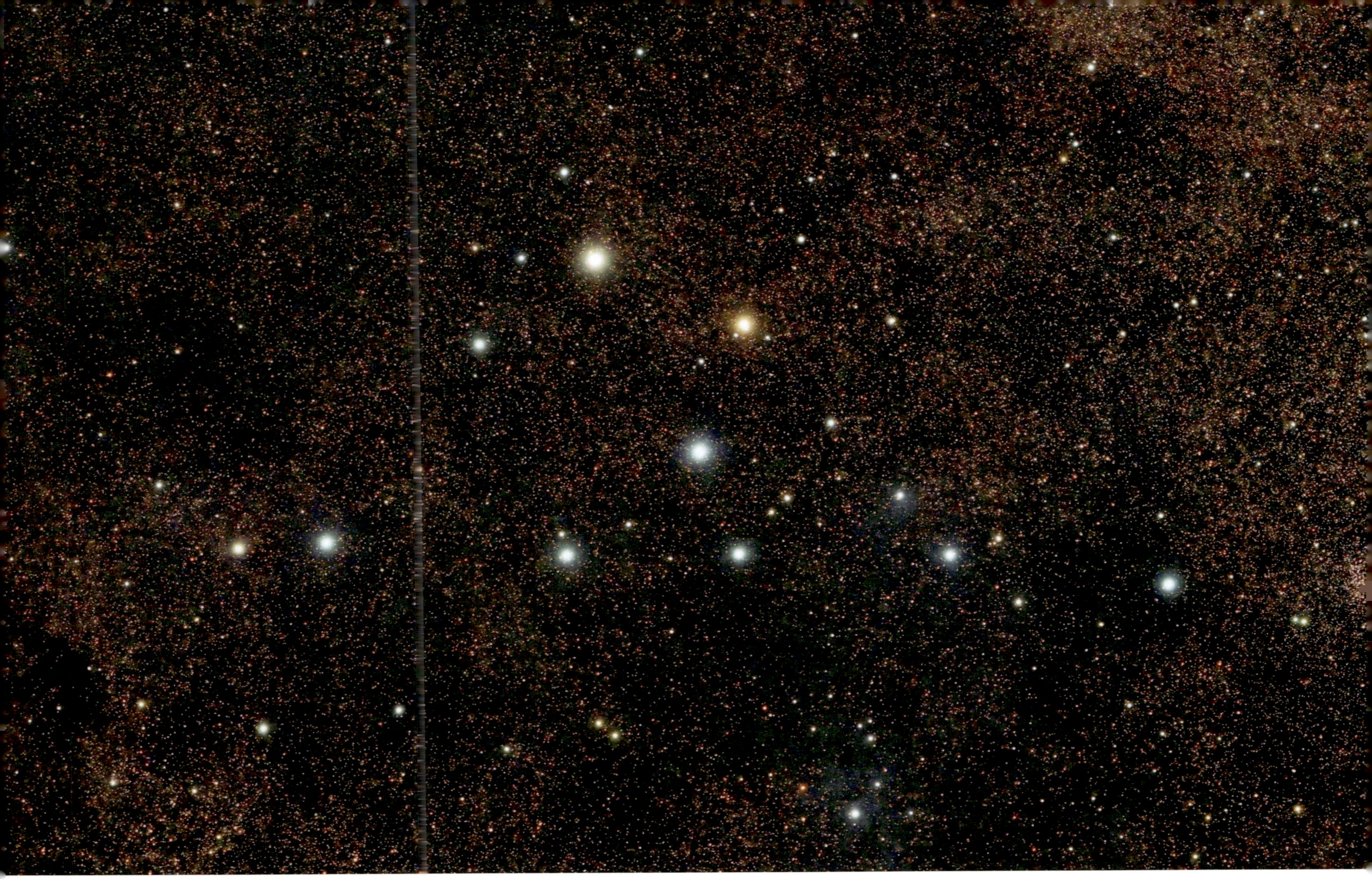

The Coathanger / Brocchi's Cluster (Collinder 399)
An asterism of 10 relatively bright stars in the constellation Vulpecula
Approximate distance from Earth: 420 light-years

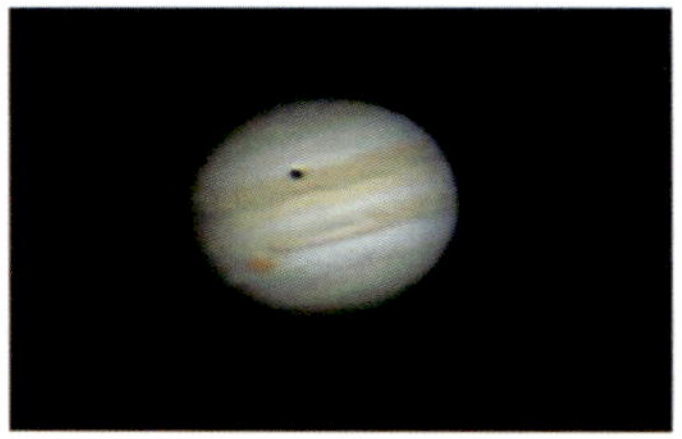

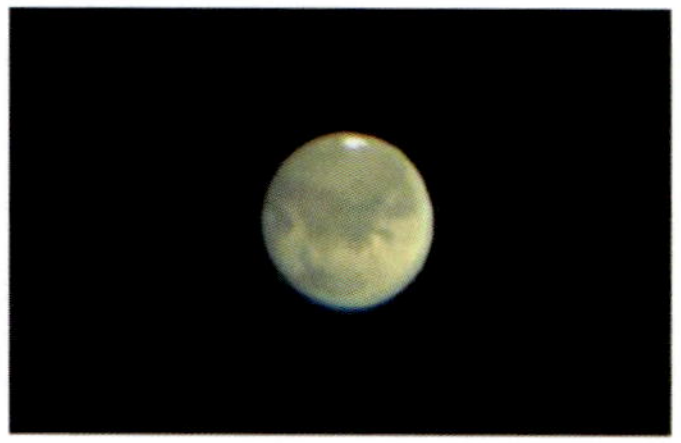

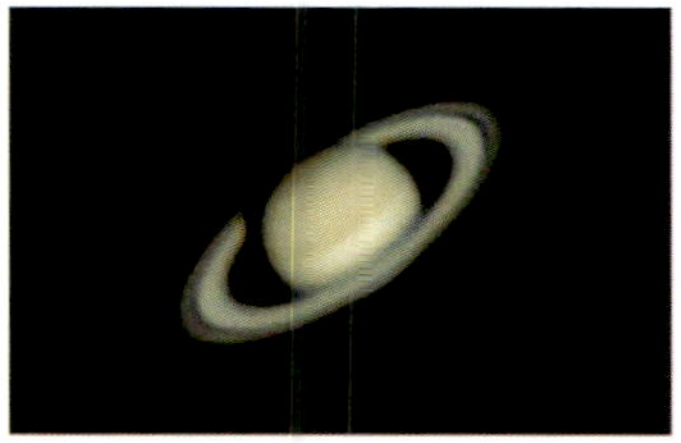

SOLAR SYSTEM IMAGES

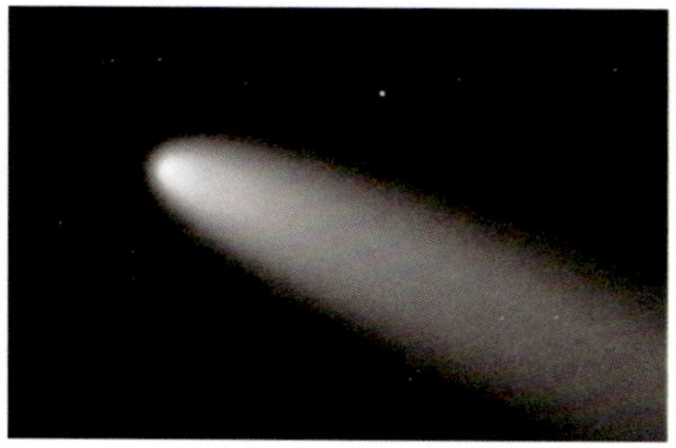

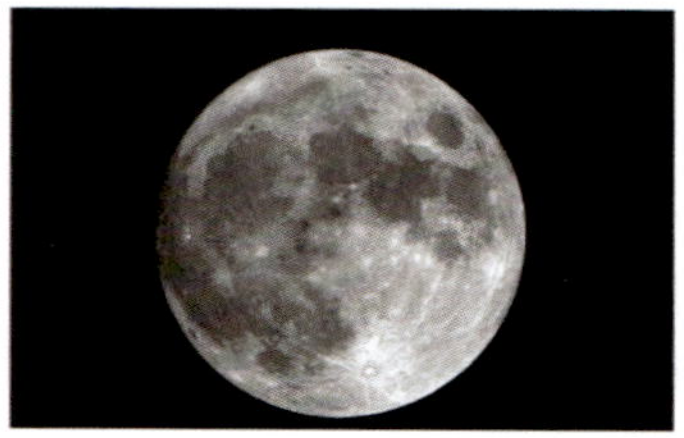

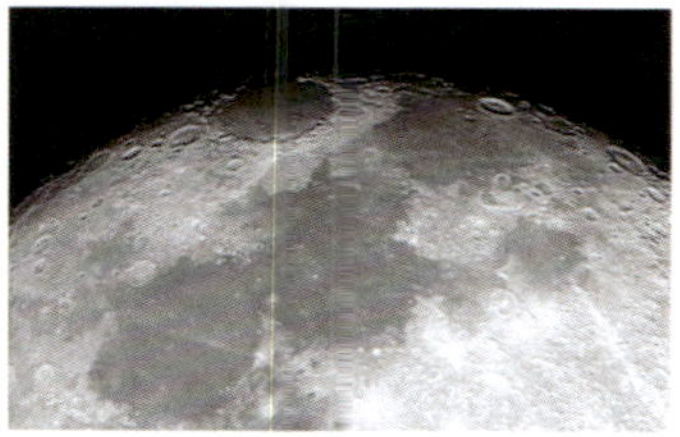

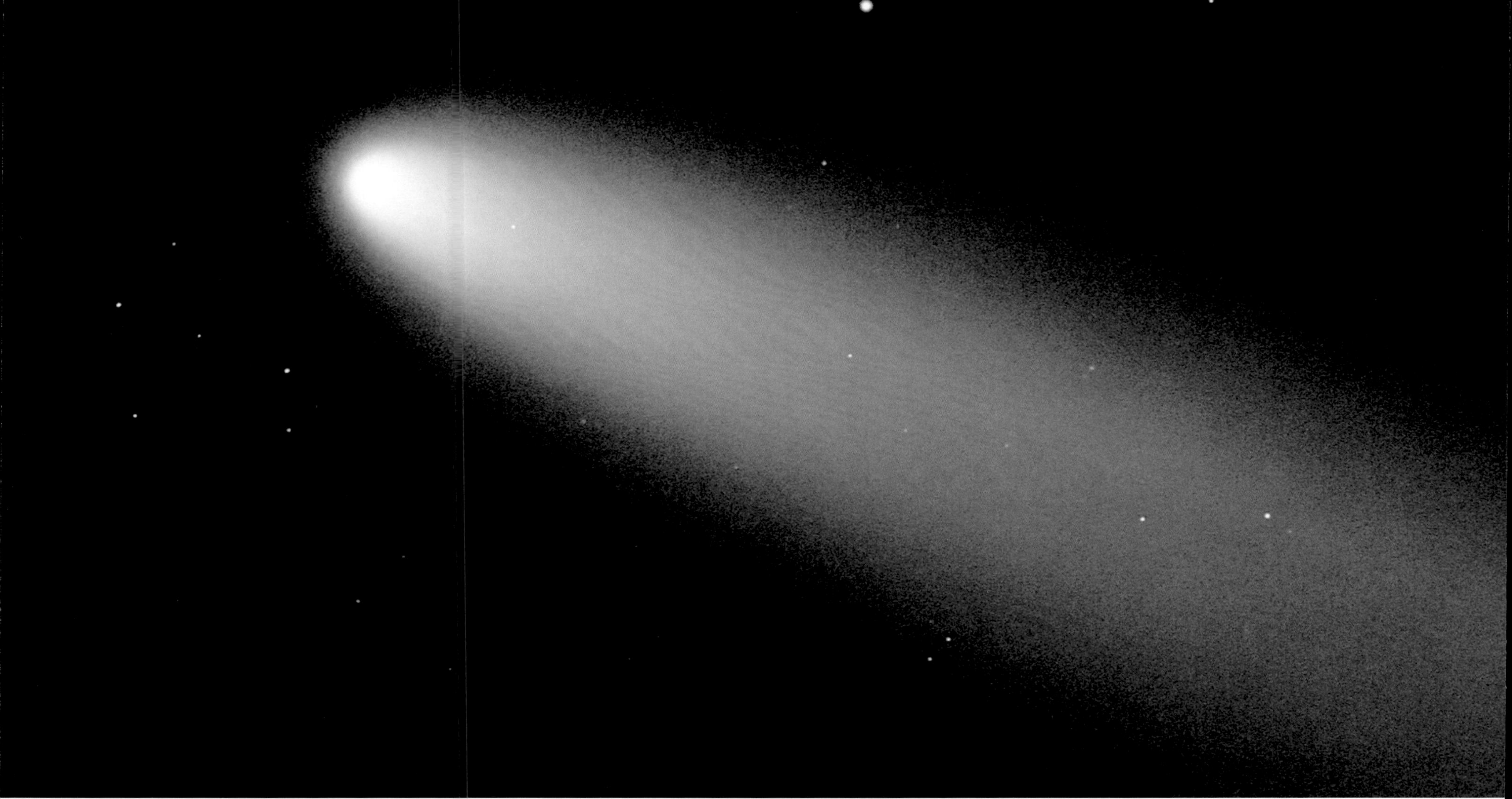

Comet C/2023 A3 (Tsuchinshan-ATLAS)

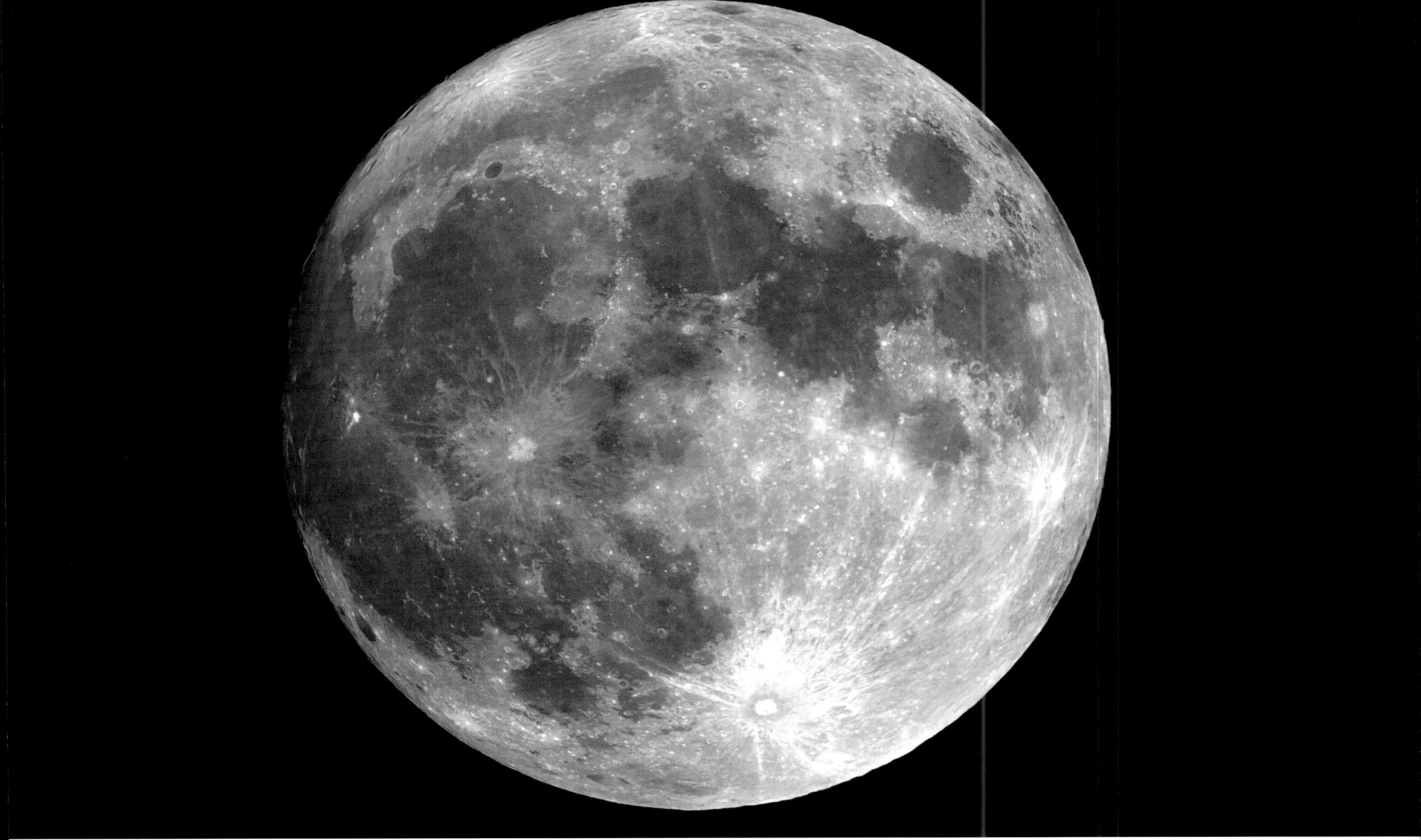

Full Moon

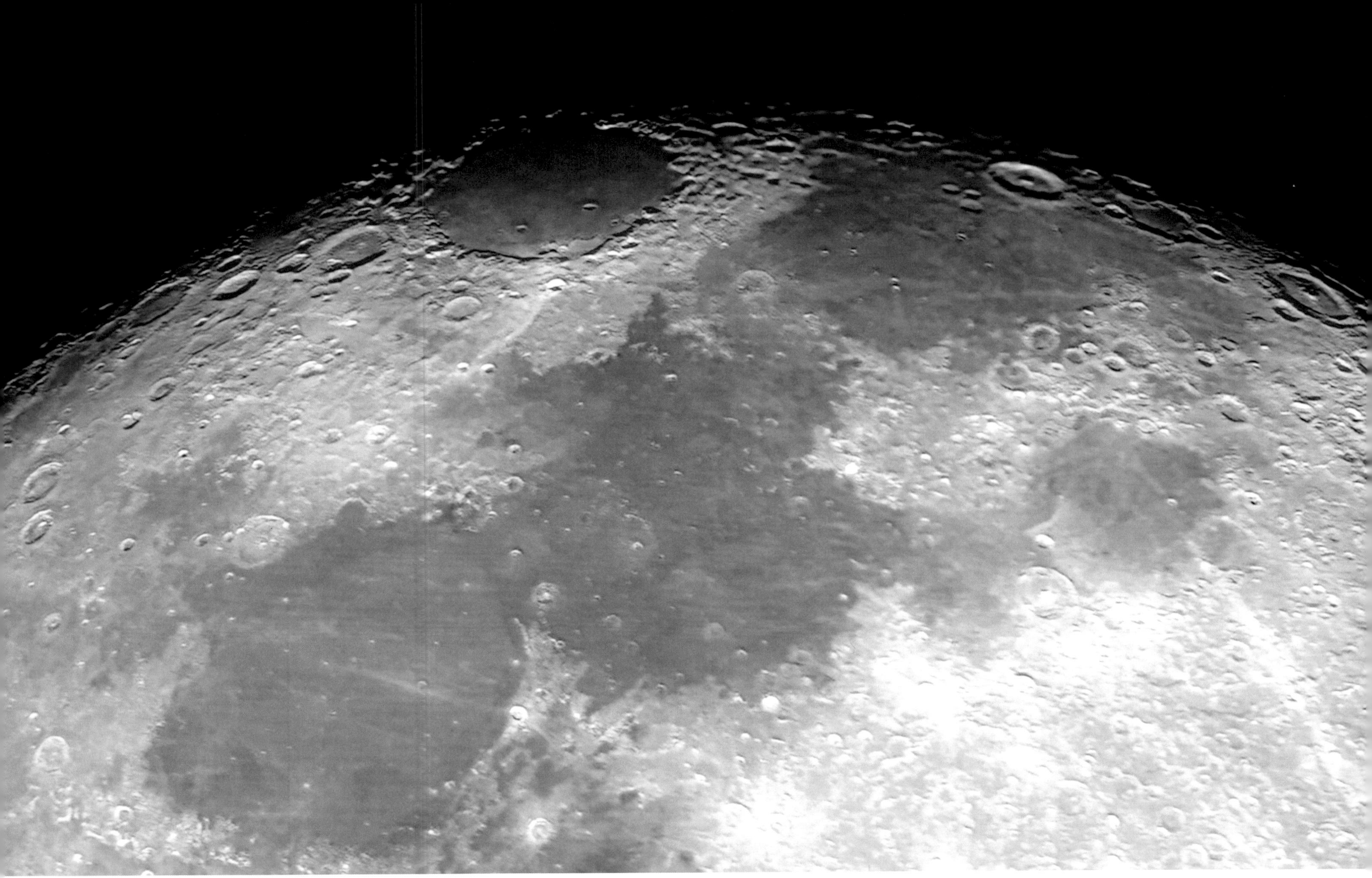

Mare Tranquillitatis (centered) / Mare Serenitatis {lower left)

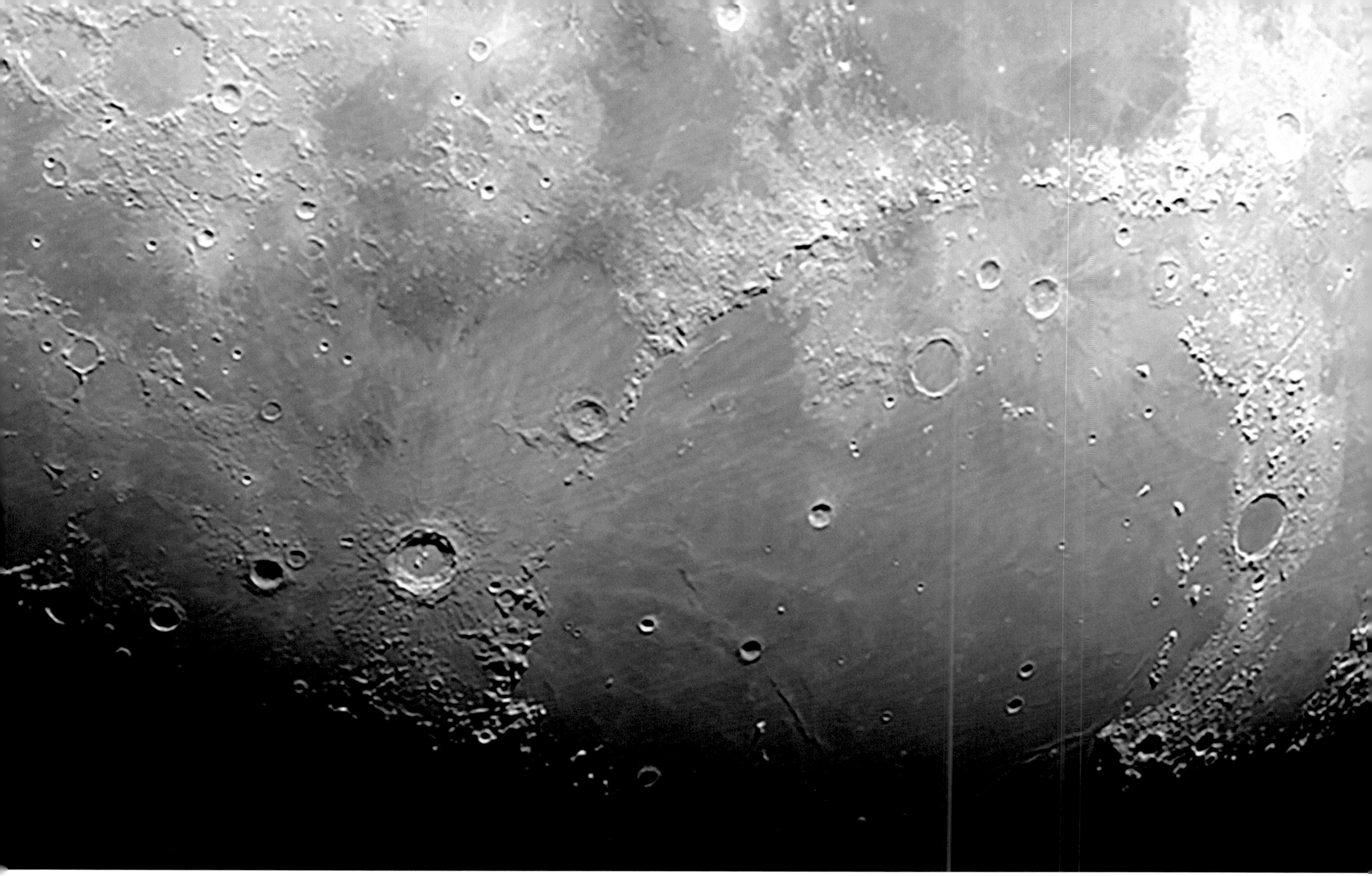

Mare Imbrium (Prominent craters Plato – Right / Copernicus – Left)

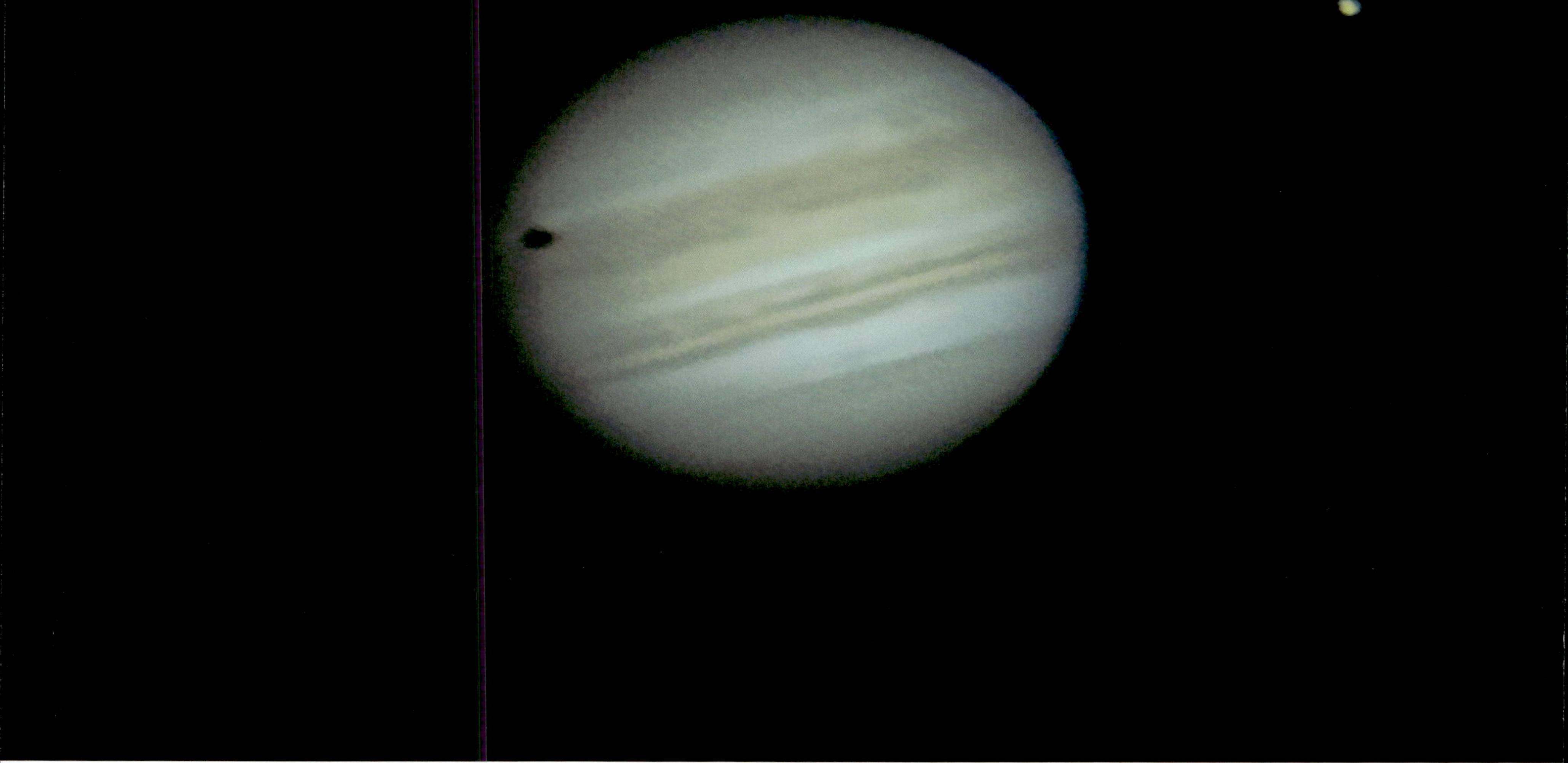

Jupiter with the moon Ganymede shadow

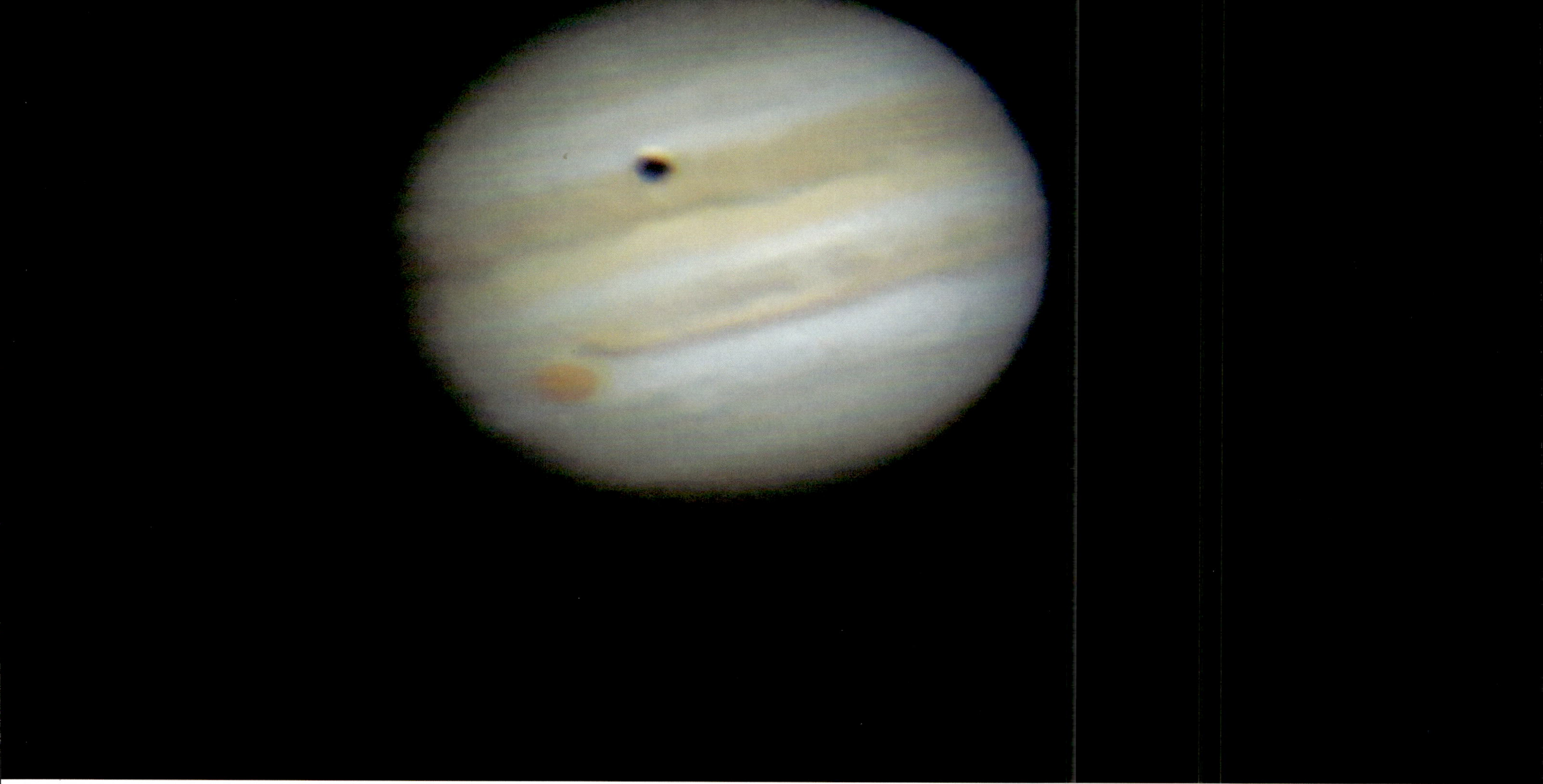

Jupiter with the moon Ganymede shadow and the Great Red Spot

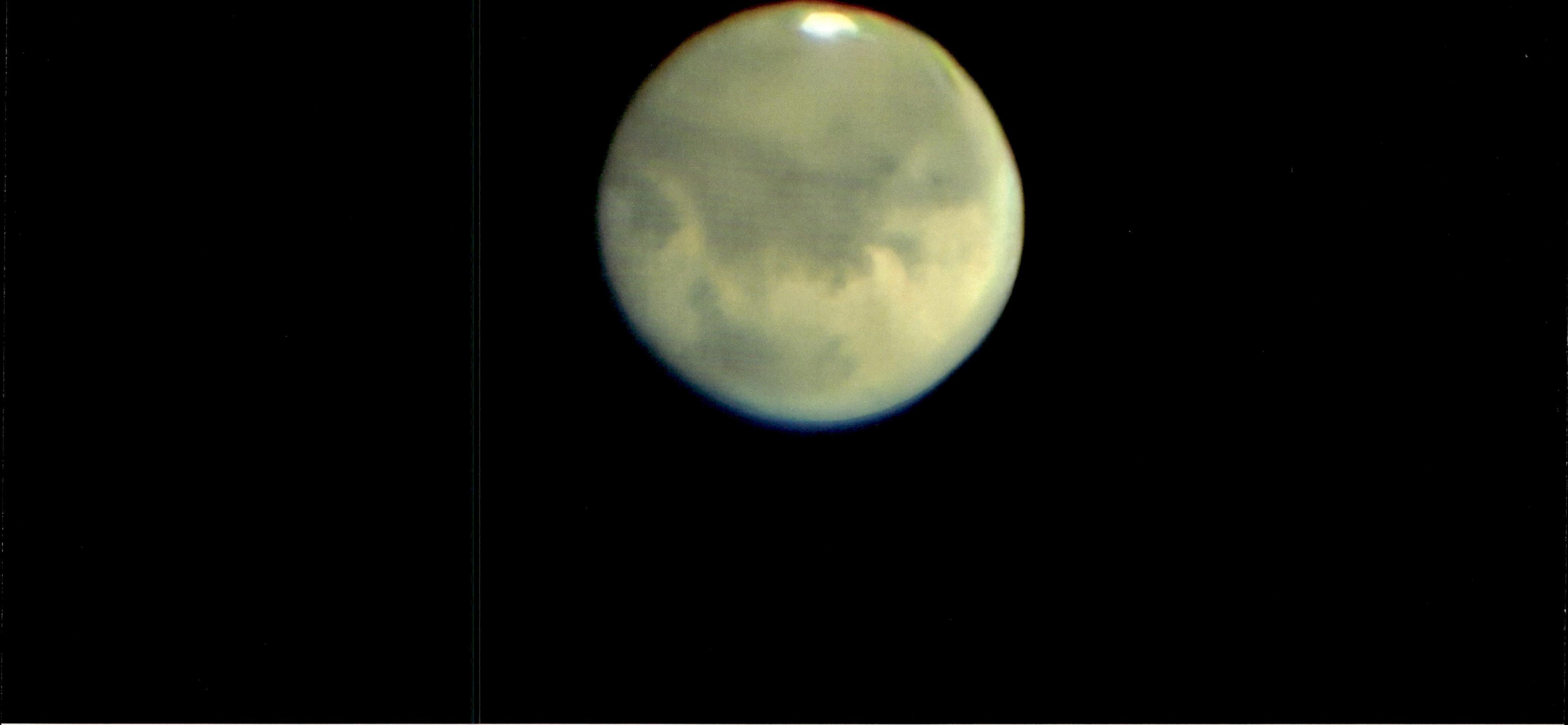

Mars with polar ice cap (white feature)

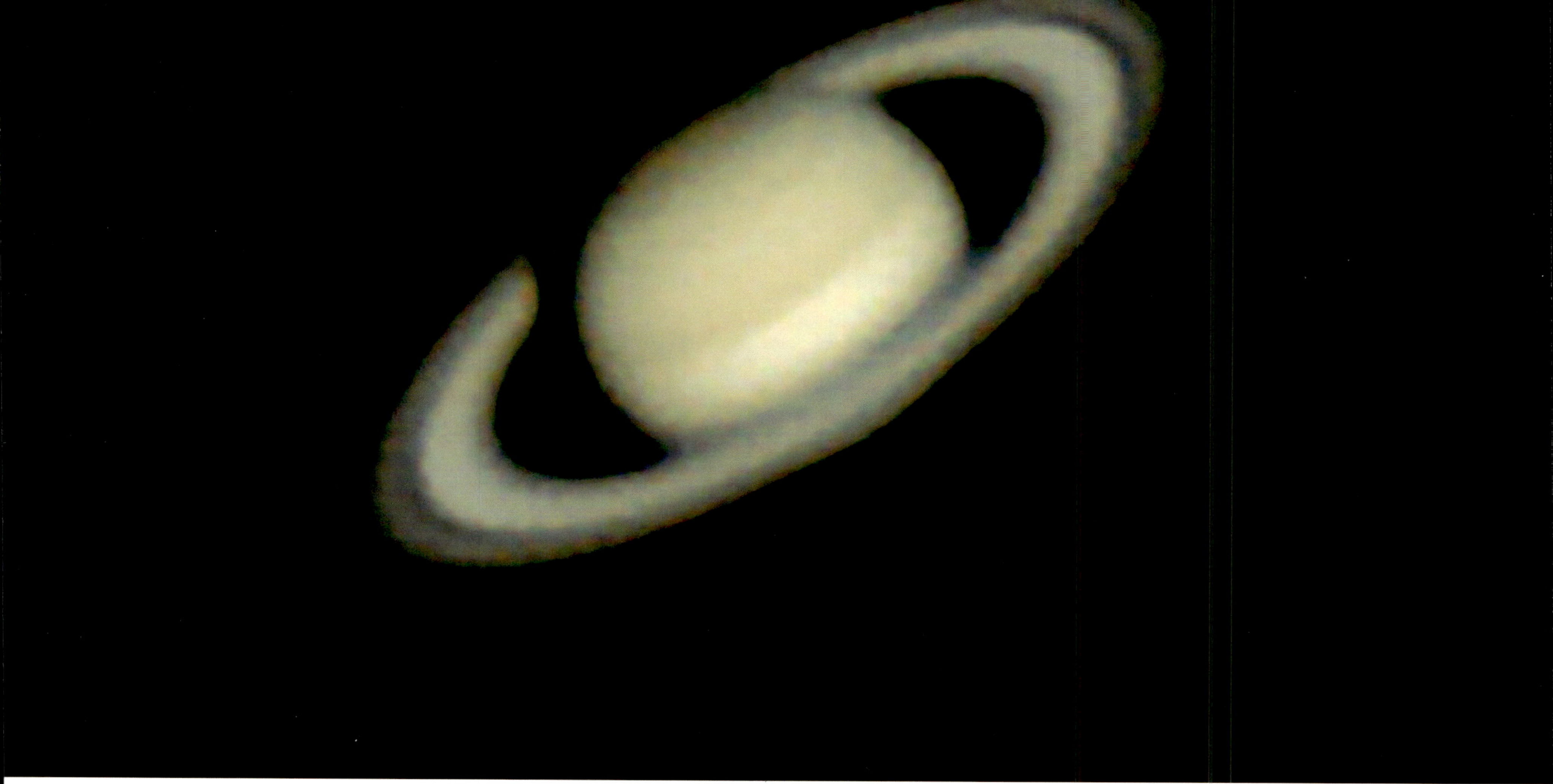

Saturn with its shadow on the rings

Appendix 1: List of Images By Constellation

Andromeda

Aquarius

Auriga

Boötes

Cancer

Canes Venatici

Canis Major

Cassiopeia

Cepheus

Coma Berenices

Corvus

Cygnus

Draco

Gemini

Hercules

Hydra

Leo

Lynx

Lyra

Monoceros

Orion

Perseus

Puppis

Sagittarius

Scorpius

Sculptor

Serpens

Taurus

Triangulum

Ursa Major

Virgo

Vulpecula

Appendix 2: Imaging Data

Common set-up data for all Deep Space images and the Comet image

Mount: Celestron: CGX-L
Filter Wheel: ZWO 7 position filter wheel
Electronic Focuser: ZWO EFA
Focusing Software: Sequence Generator Pro
Imaging Light Pollution Filter: Antlia Triband Ultra
Guide Telescope: Stellarvue SV60ED f/5.5

Guiding Camera: ZWO 174 mini Guiding Software:
Guiding Software: PHD Guiding 2 software
Capture Software: Sequence Generator Pro
Stacking Software: PixInsight
Processing Software: PixInsight and PhotoShop

Common set-up data for all Moon and Planet images

Mount: Celestron: CGX-L
Filter Wheel: ZWO 7 position filter wheel
Electronic Focuser: ZWO EFA
Focusing Software: Sequence Generator Pro

Imaging Light Pollution Filter: Antlia Triband Ultra
Capture Software: FireCapture
Processing Software: RegiStax and PhotoShop

******************************** Image Data ********************************

Page 2: M31 (Andromeda Galaxy w/M32 & M110)
Color Palette: RGB
Exposures: 100 x 240 sec
Imaging Filters: Antlia V-Pro Luminance
Imaging Telescope: Sky-Watcher Esprit 80ED f/5 (400mm EFL)
Reducer/Flattener: Standard Sky-Watcher Flattener
Imaging Camera: ZWO 2600mc / Bin 2x2

Page 3: Outer Limits Galaxy (NGC 891)
Color Palette: RGB
Exposures: 20 x 180 sec each filter
Imaging Filters: Antlia V-Pro R/G/B filters
Imaging Telescope: Sky-Watcher 150DX f/6.2 (924mm EFL)
Reducer: Sky-Watcher 0.77 Reducer/Flattener
Imaging Camera: ZWO 2600mm / Bin 2x2

Page 4: M106
Color Palette: RGB
Exposures: 50 x 240 sec
Imaging Filters: Antlia V-Pro Luminance
Imaging Telescope: Sky-Watcher Esprit 80ED f/5 (400mm EFL)
Reducer/Flattener: Standard Sky-Watcher Flattener
Imaging Camera: ZWO 2600mc / Bin 2x2

Page 5: M63 (Sunflower Galaxy)
Color Palette: HaRGB
Exposures: 45 x 120 sec each filter
Imaging Filters: Antlia V-Pro R/G/B filters
Exposures: 45 x 120 sec (Ha filter)
Imaging Filters: Antlia Edge Ha filter
Imaging Telescope: Sky-Watcher 150DX f/6.2 (924mm EFL)
Reducer: Sky-Watcher 0.77 Reducer/Flattener
Imaging Camera: ZWO 2600mm / Bin 2x2

Page 6: M51 (Whirlpool Galaxy)

Color Palette: HaRGB
Exposures: 60 x 180 sec each filter
Imaging Filters: Antlia V-Pro R/G/B filters
Exposures: 60 x 180 sec (Ha filter)
Imaging Filters: Antlia Edge Ha filter
Imaging Telescope: Sky-Watcher 150DX f/6.2 (924mm EFL)
Reducer: Sky-Watcher 0.77 Reducer/Flattener
Imaging Camera: ZWO 2600mm / Bin 2x2

Page 7: Silver Needle Galaxy (NGC 4244)

Color Palette: RGB
Exposures: 20 x 180 sec each filter
Imaging Filters: Antlia V-Pro R/G/B filters
Imaging Telescope: Sky-Watcher 150DX f/6.2 (924mm EFL)
Reducer: Sky-Watcher 0.77 Reducer/Flattener
Imaging Camera: ZWO 2600mm / Bin 2x2

Page 8: Fireworks Galaxy (NGC 6946)

Color Palette: HaRGB
Exposures: 30 x 120 sec each filter
Imaging Filters: Antlia V-Pro R/G/B filters
Exposures: 30 x 120 sec (Ha filter)
Imaging Filters: Antlia Edge Ha filter
Imaging Telescope: Sky-Watcher 150DX f/5.2 (780mm EFL)
Reducer/Flattener: Starizona Apex ED-L (0.65)
Imaging Camera: ZWO 2600mm / Bin 2x2

Page 9: M64 (Black Eye Galaxy)

Color Palette: HaRGB
Exposures: 45 x 120 sec each filter
Imaging Filters: Antlia V-Pro R/G/B filters
Exposures: 30 x 180 sec (Ha filter)
Imaging Filters: Antlia Edge Ha filter
Imaging Telescope: Sky-Watcher 150DX f/6.2 (924mm EFL)
Reducer: Sky-Watcher 0.77 Reducer/Flattener
Imaging Camera: ZWO 2600mm / Bin 2x2

Page 10: Needle Galaxy (NGC 4565)

Color Palette: RGB
Exposures: 20 x 180 sec each filter
Imaging Filters: Antlia V-Pro R/G/B filters
Imaging Telescope: Sky-Watcher 150DX f/6.2 (924mm EFL)
Reducer: Sky-Watcher 0.77 Reducer/Flattener
Imaging Camera: ZWO 2600mm / Bin 2x2

Page 11: Antennae Galaxies (NGC 4038/4039)

Color Palette: RGB
Exposures: 30 x 180 sec each filter
Imaging Filters: Antlia Edge Ha/O3/S2 filters
Imaging Telescope: Sky-Watcher 150DX f/6.2 (924mm EFL)
Reducer: Sky-Watcher 0.77 Reducer/Flattener
Imaging Camera: ZWO 2600mm / Bin 2x2

Page 12: Splinter Galaxy (NGC 5907)

Color Palette: RGB
Exposures: 20 x 120 sec each filter
Imaging Filters: Antlia V-Pro R/G/B filters
Imaging Telescope: Sky-Watcher 150DX f/6.2 (924mm EFL)
Reducer: Sky-Watcher 0.77 Reducer/Flattener
Imaging Camera: ZWO 2600mm / Bin 2x2

Page 13: M83 (Southern Pinwheel Galaxy)

Color Palette: RGB
Exposures: 16 x 180 sec each filter
Imaging Filters: Antlia V-Pro R/G/B filters
Imaging Telescope: Sky-Watcher 150DX f/6.2 (924mm EFL)
Reducer: Sky-Watcher 0.77 Reducer/Flattener
Imaging Camera: ZWO 2600mm / Bin 2x2

Page 14: Hamburger Galaxy (NGC 3628)
Color Palette: HaRGB
Exposures: 30 x 120 sec each filter
Imaging Filters: Antlia V-Pro R/G/B filters
Exposures: 45 x 240 sec (Ha filter)
Imaging Filters: Antlia Edge Ha filter
Imaging Telescope: Sky-Watcher 150DX f/6.2 (924mm EFL)
Reducer: Sky-Watcher 0.77 Reducer/Flattener
Imaging Camera: ZWO 2600mm / Bin 2x2

Page 15: M95 / M96
Color Palette: RGB
Exposures: 20 x 60 sec each filter
Imaging Filters: Antlia V-Pro R/G/B filters
Imaging Telescope: Sky-Watcher 150DX f/6.2 (924mm EFL)
Reducer: Sky-Watcher 0.77 Reducer/Flattener
Imaging Camera: ZWO 2600mm / Bin 2x2

Page 16: Sculptor / Silver Coin Galaxy (NGC 253)
Color Palette: RGB
Exposures: 24 x 300 sec each filter
Imaging Filters: Antlia V-Pro Luminance
Imaging Telescope: Sky-Watcher 150DX f/6.2 (924mm EFL)
Reducer: Sky-Watcher 0.77 Reducer/Flattener
Imaging Camera: ZWO 2600mc / Bin 2x2

Page 17: Messier 33 (Triangulum Galaxy)
Color Palette: RGB
Exposures: 150 x 180 sec each filter
Imaging Filters: Antlia V-Pro Luminance
Imaging Telescope: Sky-Watcher Esprit 80ED f/5 (400mm EFL)
Reducer/Flattener: Standard Sky-Watcher Flattener
Imaging Camera: ZWO 2600mc / Bin 2x2

Page 18: M81 (Bode's Galaxy)
Color Palette: RGB
Exposures: 75 x 180 sec each filter
Imaging Filters: Antlia V-Pro R/G/B filters
Imaging Telescope: Sky-Watcher 150DX f/6.2 (924mm EFL)
Reducer: Sky-Watcher 0.77 Reducer/Flattener
Imaging Camera: ZWO 2600mm / Bin 2x2

Page 19: M82 (Cigar Galaxy)
Color Palette: HaRGB
Exposures: 75 x 180 sec each filter
Imaging Filters: Antlia V-Pro R/G/B filters
Exposures: 30 x 120 sec + 45 x 240 sec (Ha filter)
Imaging Filters: Antlia Edge Ha filter
Imaging Telescope: Sky-Watcher 150DX f/6.2 (924mm EFL)
Reducer: Sky-Watcher 0.77 Reducer/Flattener
Imaging Camera: ZWO 2600mm / Bin 2x2

Page 20: Arp 214 Galaxy (NGC 3718)
Color Palette: RGB
Exposures: 30 x 180 sec each filter
Imaging Filters: Antlia V-Pro R/G/B filters
Imaging Telescope: Sky-Watcher 150DX f/6.2 (924mm EFL)
Reducer: Sky-Watcher 0.77 Reducer/Flattener
Imaging Camera: ZWO 2600mm / Bin 2x2

Page 21: M108 (Surfboard Galaxy)
Color Palette: HaRGB
Exposures: 25 x 120 sec each filter
Imaging Filters: Antlia V-Pro R/G/B filters
Exposures: 25 x 120 sec (Ha filter)
Imaging Filters: Antlia Edge Ha filter
Imaging Telescope: Sky-Watcher 150DX f/6.2 (924mm EFL)
Reducer: Sky-Watcher 0.77 Reducer/Flattener
Imaging Camera: ZWO 2600mm / Bin 2x2

Page 22: M109
Color Palette: RGB
Exposures: 60 x 120 sec each filter
Imaging Filters: Antlia V-Pro R/G/B filters
Imaging Telescope: Sky-Watcher 150DX f/6.2 (924mm EFL)
Reducer: Sky-Watcher 0.77 Reducer/Flattener
Imaging Camera: ZWO 2600mm / Bin 2x2

Page 23: M101 (Pinwheel Galaxy)
Color Palette: RGB
Exposures: 25 x 180 sec each filter
Imaging Filters: Antlia V-Pro R/G/B filters
Imaging Telescope: Sky-Watcher 150DX f/6.2 (924mm EFL)
Reducer: Sky-Watcher 0.77 Reducer/Flattener
Imaging Camera: ZWO 2600mm / Bin 2x2

Page 24: M104 (Sombrero Galaxy)
Color Palette: RGB
Exposures: 25 x 180 sec each filter
Imaging Filters: Antlia V-Pro R/G/B filters
Imaging Telescope: Sky-Watcher 150DX f/6.2 (924mm EFL)
Reducer: Sky-Watcher 0.77 Reducer/Flattener
Imaging Camera: ZWO 2600mm / Bin 2x2

Page 26: Helix Nebula (NGC 7293)
Color Palette: RGB
Exposures: 60 x 180 sec each filter
Imaging Filters: Antlia V-Pro Luminance
Imaging Telescope: Sky-Watcher Esprit 80ED f/5 (400mm EFL)
Reducer/Flattener: Standard Sky-Watcher Flattener
Imaging Camera: ZWO 2600mc / Bin 2x2

Page 27: Flaming Star Nebula (IC 405)
Color Palette: RGB
Exposures: 92 x 180 sec each filter
Imaging Filters: Antlia V-Pro Luminance
Imaging Telescope: Sky-Watcher Esprit 80ED f/5 (400mm EFL)
Reducer/Flattener: Standard Sky-Watcher Flattener
Imaging Camera: ZWO 2600mc / Bin 2x2

Page 28: Tadpoles Nebula (IC 410)
Color Palette: RGB
Exposures: 80 x 180 sec each filter
Imaging Filters: Antlia V-Pro Luminance
Imaging Telescope: Sky-Watcher Esprit 80ED f/5 (400mm EFL)
Reducer/Flattener: Standard Sky-Watcher Flattener
Imaging Camera: ZWO 2600mc / Bin 2x2

Page 29: Thor's Helmet (NGC 2359)
Color Palette: HOO
Exposures: 43 x 120 sec each filter
Imaging Filters: Antlia Edge Ha/O3/S2 filters
Imaging Telescope: Sky-Watcher 150DX f/6.2 (924mm EFL)
Reducer: Sky-Watcher 0.77 Reducer/Flattener
Imaging Camera: ZWO 2600mm / Bin 2x2

Page 30: Popped Balloon Nebula (LBN 576)
Color Palette: HOO
Exposures: 20 x 240 sec each filter
Imaging Filters: Antlia Edge Ha/O3 filters
Imaging Telescope: Sky-Watcher 150DX f/6.2 (924mm EFL)
Reducer: Sky-Watcher 0.77 Reducer/Flattener
Imaging Camera: ZWO 2600mm / Bin 2x2

Page 31: Soul Nebula (IC 1848)
Color Palette: SHO
Exposures: 20 x 180 sec each filter
Imaging Filters: Antlia Edge Ha/O3/S2 filters
Imaging Telescope: Sky-Watcher Esprit 80ED f/5 (400mm EFL)
Reducer/Flattener: Standard Sky-Watcher Flattener
Imaging Camera: ZWO 2600mm / Bin 2x2

Page 32: Heart Nebula (IC 1805)
Color Palette: SHO
Exposures: 25 x 180 sec each filter
Imaging Filters: Antlia Edge Ha/O3/S2 filters
Imaging Telescope: Sky-Watcher Esprit 80ED f/5 (400mm EFL)
Reducer/Flattener: Standard Sky-Watcher Flattener
Imaging Camera: ZWO 2600mm / Bin 2x2
Processing Software: PixInsight and PhotoShop

Page 33: Melotte 15 The Center of the Heart Nebula
Color Palette: SHO
Exposures: 25 x 180 sec each filter
Imaging Filters: Antlia Edge Ha/O3/S2 filters
Imaging Telescope: Sky-Watcher Esprit 80ED f/5 (400mm EFL)
Reducer/Flattener: Standard Sky-Watcher Flattener
Imaging Camera: ZWO 2600mm / Bin 2x2

Page 34: Lobster Claw Nebula Mosaic (Sh2-157)
Color Palette: SHO – 2 panel mosaic
Exposures: 40 x 120 sec each filter
Imaging Filters: Antlia Edge Ha/O3/S2 filters
Imaging Telescope: Sky-Watcher 150DX f/6.2 (924mm EFL)
Reducer: Sky-Watcher 0.77 Reducer/Flattener
Imaging Camera: ZWO 2600mm / Bin 2x2

Page 35: Phantom of the Opera Nebula (Sh2-173)
Color Palette: HOO
Exposures: 15 x 120 sec each filter
Imaging Filters: Antlia Edge Ha/O3 filters
Imaging Telescope: Sky-Watcher Esprit 80ED f/5 (400mm EFL)
Reducer/Flattener: Standard Sky-Watcher Flattener
Imaging Camera: ZWO 2600mc / Bin 2x2

Page 36: Ghost of Cassiopeiae (IC 63)
Color Palette: SHO
Exposures: 30 x 240 sec each filter
Imaging Filters: Antlia Edge Ha/O3/S2 filters
Imaging Telescope: Sky-Watcher 150DX f/6 2 (924mm EFL)
Reducer: Sky-Watcher 0.77 Reducer/Flattener
Imaging Camera: ZWO 2600mm / Bin 2x2

Page 37: Bubble Nebula (NGC 7635)
Color Palette: SHO
Exposures: 20 x 240 sec each filter
Imaging Filters: Antlia Edge Ha/O3/S2 filters
Imaging Telescope: Sky-Watcher 150DX f/6 2 (924mm EFL)
Reducer: Sky-Watcher 0.77 Reducer/Flattener
Imaging Camera: ZWO 2600mm / Bin 2x2

Page 38: Pac-Man Nebula (NGC 281)
Color Palette: HOO
Exposures: 20 x 180 sec each filter
Imaging Filters: Antlia Edge Ha/O3 filters
Imaging Telescope: Sky-Watcher 150DX f/6.2 (924mm EFL)
Reducer: Sky-Watcher 0.77 Reducer/Flattener
Imaging Camera: ZWO 2600mm / Bin 2x2

Page 39: Iris Nebula (NGC 7023)
Color Palette: RGB
Exposures: 40 x 120 sec each filter
Imaging Filters: Antlia Edge Ha/O3/S2 filters
Imaging Telescope: Sky-Watcher 150DX f/6.2 (924mm EFL)
Reducer: Sky-Watcher 0.77 Reducer/Flattener
Imaging Camera: ZWO 2600mm / Bin 2x2

Page 40: Wizard Nebula (Sh2-142)
Color Palette: SHO
Exposures: 20 x 240 sec each filter
Imaging Filters: Antlia Edge Ha/O3/S2 filters
Imaging Telescope: Sky-Watcher 150DX f/6.2 (924mm EFL)
Reducer: Sky-Watcher 0.77 Reducer/Flattener
Imaging Camera: ZWO 2600mm / Bin 2x2

Page 41: Elephant Trunk Nebula (IC 1396)
Color Palette: SHO
Exposures: 40 x 180 sec each filter
Imaging Filters: Antlia Edge Ha/O3/S2 filters
Imaging Telescope: Sky-Watcher 150DX f/6.2 (924mm EFL)
Reducer: Sky-Watcher 0.77 Reducer/Flattener
Imaging Camera: ZWO 2600mm / Bin 2x2

Page 42: Tulip Nebula (Sh2-101)
Color Palette: SHO
Exposures: 25 x 180 sec each filter
Imaging Filters: Antlia Edge Ha/O3/S2 filters
Imaging Telescope: Sky-Watcher 150DX f/6.2 (924mm EFL)
Reducer: Sky-Watcher 0.77 Reducer/Flattener
Imaging Camera: ZWO 2600mm / Bin 2x2

Page 43: Soap Bubble Nebula
Color Palette: HOO
Exposures: 25 x 240 sec each filter
Imaging Filters: Antlia Edge Ha/O3 filters
Imaging Telescope: Sky-Watcher 150DX f/6.2 (924mm EFL)
Reducer: Sky-Watcher 0.77 Reducer/Flattener
Imaging Camera: ZWO 2600mm / Bin 2x2

Page 44: Veil Nebula Full (Cygnus Loop)
Color Palette: RGB
Exposures: 40 x 180 sec each filter
Imaging Filters: Antlia V-Pro Luminance
Imaging Telescope: Sky-Watcher Esprit 80ED f/5 (400mm EFL)
Reducer/Flattener: Standard Sky-Watcher Flattener
Imaging Camera: ZWO 2600mc / Bin 2x2

Page 45: Veil Nebula East (NGC 6992 / 6995)
Color Palette: HOO
Exposures: 40 x 60 sec each filter
Imaging Filters: Antlia Edge Ha/O3 filters
Imaging Telescope: Sky-Watcher 150DX f/6.2 (924mm EFL)
Reducer: Sky-Watcher 0.77 Reducer/Flattener
Imaging Camera: ZWO 2600mM / Bin 2x2

Page 46: Veil Nebula West (Witch's Broom NGC 6960)
Color Palette: HOO
Exposures: 30 x 60 sec each filter
Imaging Filters: Antlia Edge Ha/O3 filters
Imaging Telescope: Sky-Watcher 150DX f/6.2 (924mm EFL)
Reducer: Sky-Watcher 0.77 Reducer/Flattener
Imaging Camera: ZWO 2600mm / Bin 2x2

Page 47: Veil Nebula (Pickering's Triangle)
Color Palette: HOO
Exposures: 40 x 60 sec each filter
Imaging Filters: Antlia Edge Ha/O3 filters
Imaging Telescope: Sky-Watcher 150DX f/6.2 (924mm EFL)
Reducer: Sky-Watcher 0.77 Reducer/Flattener
Imaging Camera: ZWO 2600mM / Bin 2x2

Page 48: Crescent Nebula (NGC 6888)
Color Palette: HOO
Exposures: 25 x 240 sec each filter
Imaging Filters: Antlia Edge Ha/O3 filters
Imaging Telescope: Sky-Watcher 150DX f/6.2 (924mm EFL)
Reducer: Sky-Watcher 0.77 Reducer/Flattener
Imaging Camera: ZWO 2600mm / Bin 2x2

Page 49: North American & Pelican Nebula Mosaic
Color Palette: SHO – 2 panel mosaic
Exposures: 20 x 180 sec each filter
Imaging Filters: Antlia Edge Ha/O3/S2 filters
Imaging Telescope: Sky-Watcher Esprit 80ED f/5 (400mm EFL)
Reducer/Flattener: Standard Sky-Watcher Flattener
Imaging Camera: ZWO 2600mm / Bin 2x2

Page 50: North American Nebula (NGC 7000)
Color Palette: SHO
Exposures: 20 x 180 sec each filter
Imaging Filters: Antlia Edge Ha/O3/S2 filters
Imaging Telescope: Sky-Watcher Esprit 80ED f/5 (400mm EFL)
Reducer/Flattener: Standard Sky-Watcher Flattener
Imaging Camera: ZWO 2600mm / Bin 2x2

Page 51: Pelican Nebula (IC 5070)
Color Palette: SHO
Exposures: 20 x 180 sec each filter
Imaging Filters: Antlia Edge Ha/O3/S2 filters
Imaging Telescope: Sky-Watcher Esprit 80ED f/5 (400mm EFL)
Reducer/Flattener: Standard Sky-Watcher Flattener
Imaging Camera: ZWO 2600mm / Bin 2x2

Page 52: Cocoon Nebula — (IC 5146)
Color Palette: HOO
Exposures: 30 x 120 sec each filter
Imaging Filters: Antlia Edge Ha/O3 filters
Imaging Telescope: Sky-Watcher 150DX f/6 2 (924mm EFL)
Reducer: Sky-Watcher 0.77 Reducer/Flattener
Imaging Camera: ZWO 2600mm / Bin 2x2

Page 53: Flying Dragon Nebula (Sh2-114)
Color Palette: HSS
Exposures: 25 x 180 sec each filter
Imaging Filters: Antlia Edge Ha/S2 filters
Imaging Telescope: Sky-Watcher Esprit 80ED f/5 (400mm EFL)
Reducer/Flattener: Standard Sky-Watcher Flattener
Imaging Camera: ZWO 2600mm / Bin 2x2

Page 54: Cat's Eye Nebula (NGC 6543)
Color Palette: HOO
Exposures: 60 x 30 sec each filter
Imaging Filters: Antlia Edge Ha/O3 filters
Exposures: 40 x 180 sec each filter
Imaging Filters: Antlia Edge Ha/O3 filters
Imaging Telescope: Sky-Watcher 150DX f/6.2 (924mm EFL)
Reducer: Sky-Watcher 0.77 Reducer/Flattener
Imaging Camera: ZWO 2600mm / Bin 2x2

Page 55: Jellyfish Nebula (IC 443)
Color Palette: RGB
Exposures: 60 x 180 sec
Imaging Filters: Antlia V-Pro Luminance
Imaging Telescope: Sky-Watcher Esprit 80ED f/5 (400mm EFL)
Reducer/Flattener: Standard Sky-Watcher Flattener
Imaging Camera: ZWO 2600mc / Bin 2x2

Page 56: Medusa
Color Palette: HOO
Exposures: 18 x 120 sec
Imaging Filters: Antlia Edge Ha/O3/S2 filters
Imaging Telescope: Sky-Watcher 150DX f/6.2 (924mm EFL)
Reducer: Sky-Watcher 0.77 Reducer/Flattener
Imaging Camera: ZWO 2600mm / Bin 2x2

Page 57: Eskimo Nebula (NGC 2392)
Color Palette: RGB
Exposures: 60 x 10 sec
Imaging Filters: Antlia V-Pro Luminance
Imaging Telescope: Sky-Watcher 150DX f/6.2 (924mm EFL)
Reducer: Sky-Watcher 0.77 Reducer/Flattener
Imaging Camera: ZWO 2600mc / Bin 2x2

Page 58: Headphone Nebula — (PK 164+31.1)
Color Palette: HOO
Exposures: 50 x 180 sec each filter
Imaging Filters: Antlia Edge Ha/O3 filters
Imaging Telescope: Sky-Watcher 150DX f/6.2 (924mm EFL)
Reducer: Sky-Watcher 0.77 Reducer/Flattener
Imaging Camera: ZWO 2600mm / Bin 2x2

Page 59: M57 (Ring Nebula)
Color Palette: HOO
Exposures: 40 x 30 sec each filter
Imaging Filters: Antlia Edge Ha/O3 filters
Imaging Telescope: Sky-Watcher 150DX f/6.2 (924mm EFL)
Reducer: Sky-Watcher 0.77 Reducer/Flattener
Imaging Camera: ZWO 2600mm / Bin 2x2

Page 60: Christmas Tree Nebula / Cluster (NGC 2264)
Color Palette: RGB
Exposures: 60 x 180 sec
Imaging Filters: Antlia V-Pro Luminance
Imaging Telescope: Sky-Watcher Esprit 80ED f/5 (400mm EFL)
Reducer/Flattener: Standard Sky-Watcher Flattener
Imaging Camera: ZWO 2600mc / Bin 2x2

Page 61: Cone Nebula (NGC 2264)
Color Palette: RGB
Exposures: 60 x 180 sec
Imaging Filters: Antlia V-Pro Luminance
Imaging Telescope: Sky-Watcher Esprit 80ED f/5 (400mm EFL)
Reducer/Flattener: Standard Sky-Watcher Flattener
Imaging Camera: ZWO 2600mc / Bin 2x2

Page 62: Rosette Nebula (NGC 2237)
Color Palette: RGB
Exposures: 80 x 180 sec
Imaging Filters: Antlia V-Pro Luminance
Imaging Telescope: Sky-Watcher Esprit 80ED f/5 (400mm EFL)
Reducer/Flattener: Standard Sky-Watcher Flattener
Imaging Camera: ZWO 2600mc / Bin 2x2

Page 63: Monkey Head Nebula (NGC 2174)
Color Palette: RGB
Exposures: 70 x 180 sec
Imaging Filters: Antlia V-Pro Luminance
Imaging Telescope: Sky-Watcher Esprit 80ED f/5 (400mm EFL)
Reducer/Flattener: Standard Sky-Watcher Flattener
Imaging Camera: ZWO 2600mc / Bin 2x2

Page 64: M42 (Orion Nebula)
Color Palette: RGB
Exposures: 60 x 120 sec each filter
Imaging Filters: Antlia V-Pro Luminance
Imaging Telescope: Sky-Watcher Esprit 80ED f/5 (400mm EFL)
Reducer/Flattener: Standard Sky-Watcher Flattener
Imaging Camera: ZWO 2600mc / Bin 2x2

Page 65: Running Man Nebula (NGC 1977)
Color Palette: RGB
Exposures: 10 x 60 sec each filter
Imaging Filters: Antlia V-Pro R/G/B filters
Imaging Telescope: Sky-Watcher Esprit 80ED f/5 (400mm EFL)
Reducer/Flattener: Standard Sky-Watcher Flattener
Imaging Camera: ZWO 2600mm / Bin 2x2

Page 66: Flame Nebula (NGC 2024)
Color Palette: RGB
Exposures: 180 x 120 sec each filter
Imaging Filters: Antlia V-Pro Luminance
Imaging Telescope: Sky-Watcher Esprit 80ED f/3.25 (260mm EFL)
Reducer/Flattener: Starizona Apex ED-S (0.65)
Imaging Camera: ZWO 2600mc / Bin 2x2

Page 67: Horsehead Nebula (B33)
Color Palette: HaRGB
Exposures: 30 x 60 sec each filter
Imaging Filters: Antlia V-Pro R/G/B filters
Exposures: 30 x 60 sec (Ha filter)
Imaging Filters: Antlia Edge Ha filter
Imaging Telescope: Sky-Watcher 150DX f/5.2 (924mm EFL)
Reducer: Sky-Watcher 0.77 Reducer/Flattener
Imaging Camera: ZWO 2600mm / Bin 2x2

Page 68: California Nebula (NGC 1499)
Color Palette: HOO
Exposures: 20 x 180 sec each filter
Imaging Filters: Antlia Edge Ha/O3 filters
Imaging Telescope: Sky-Watcher Esprit 80ED f/5 (400mm EFL)
Reducer/Flattener: Standard Sky-Watcher Flattener
Imaging Camera: ZWO 2600mm / Bin 2x2

Page 69: M17 (Swan/Omega Nebula)
Color Palette: SHO
Exposures: 20 x 60 sec each filter
Imaging Filters: Antlia Edge Ha/O3/S2 filters
Imaging Telescope: Sky-Watcher 150DX f/5.2 (780mm EFL)
Reducer/Flattener: Starizona Apex ED-L (0.65)
Imaging Camera: ZWO 2600mm / Bin 2x2

Page 70: M20 (Trifid Nebula)
Color Palette: RGB
Exposures: 55 x 120 sec each filter
Imaging Filters: Antlia V-Pro R/G filters
Exposures: 21 x 240 sec
Imaging Filters: Antlia V-Pro B filter
Imaging Telescope: Sky-Watcher 150DX f/6.2 (924mm EFL)
Reducer: Sky-Watcher 0.77 Reducer/Flattener
Imaging Camera: ZWO 2600mm / Bin 2x2

Page 71: M8 (Lagoon Nebula)
Color Palette: RGB
Exposures: 20 x 180 sec each filter
Imaging Filters: Antlia V-Pro R/G/B filters
Imaging Telescope: Sky-Watcher Esprit 80ED f/5 (400mm EFL)
Reducer/Flattener: Standard Sky-Watcher Flattener
Imaging Camera: ZWO 2600mm / Bin 2x2

Page 72: Bug / Butterfly Nebula — (NGC 6302)
Color Palette: HOO
Exposures: 30 x 60 sec each filter
Imaging Filters: Antlia Edge Ha/O3 filters
Imaging Telescope: Sky-Watcher 150DX f/6.2 (924mm EFL)
Reducer: Sky-Watcher 0.77 Reducer/Flattener
Imaging Camera: ZWO 2600mm / Bin 2x2

Page 73: M16 (Eagle Nebula – "Pillars of Creation")
Color Palette: SHO
Exposures: 20 x 180 sec each filter
Imaging Filters: Antlia Edge Ha/O3/S2 filters
Imaging Telescope: Sky-Watcher Esprit 80ED f/5 (400mm EFL)
Reducer/Flattener: Standard Sky-Watcher Flattener
Imaging Camera: ZWO 2600mc / Bin 2x2

Page 74: M1 (Crab Nebula)
Color Palette: SHO
Exposures: 20 x 180 sec each filter
Imaging Filters: Antlia Edge Ha/O3/S2 filters
Imaging Telescope: Sky-Watcher 150DX f/6.2 (924mm EFL)
Reducer: Sky-Watcher 0.77 Reducer/Flattener
Imaging Camera: ZWO 2600mm / Bin 2x2

Page 75: M97 (Owl Nebula)
Color Palette: HOO
Exposures: 30 x 120 sec each filter
Imaging Filters: Antlia Edge Ha/O3 filters
Imaging Telescope: Sky-Watcher 150DX f/6.2 (924mm EFL)
Reducer: Sky-Watcher 0.77 Reducer/Flattener
Imaging Camera: ZWO 2600mm / Bin 2x2

Page 76: Sh2-86 (NGC 6820 / 6823)
Color Palette: SHO
Exposures: 20 x 240 sec each filter
Imaging Filters: Antlia Edge Ha/O3/S2 filters
Imaging Telescope: Sky-Watcher 150DX f/6.2 (924mm EFL)
Reducer: Sky-Watcher 0.77 Reducer/Flattener
Imaging Camera: ZWO 2600mm / Bin 2x2

Page 77: M27 (Dumbbell Nebula)
Color Palette: HOO
Exposures: 30 x 120 sec each filter
Imaging Filters: Antlia Edge Ha/O3 filters
Imaging Telescope: Sky-Watcher 150DX f/6.2 (924mm EFL)
Reducer: Sky-Watcher 0.77 Reducer/Flattener
Imaging Camera: ZWO 2600mm / Bin 2x2

Page 79: M37
Color Palette: RGB
Exposures: 16 x 30 sec each filter
Imaging Filters: Antlia V-Pro R/G/B filters
Imaging Telescope: Sky-Watcher Esprit 80ED f/5 (400mm EFL)
Reducer/Flattener: Standard Sky-Watcher Flattener
Imaging Camera: ZWO 2600mm / Bin 2x2

Page 80: M38
Color Palette: RGB
Exposures: 16 x 30 sec each filter
Imaging Filters: Antlia V-Pro R/G/B filters
Imaging Telescope: Sky-Watcher Esprit 80ED f/5 (400mm EFL)
Reducer/Flattener: Standard Sky-Watcher Flattener
Imaging Camera: ZWO 2600mm / Bin 2x2

Page 81: M36
Color Palette: RGB
Exposures: 16 x 30 sec each filter
Imaging Filters: Antlia V-Pro R/G/B filters
Imaging Telescope: Sky-Watcher Esprit 80ED f/5 (400mm EFL)
Reducer/Flattener: Standard Sky-Watcher Flattener
Imaging Camera: ZWO 2600mm / Bin 2x2

Page 82: M3 (Globular Cluster)
Color Palette: RGB
Exposures: 60 x 30 sec each filter
Imaging Filters: Antlia V-Pro R/G/B filters
Imaging Telescope: Sky-Watcher 150DX f/6.2 (924mm EFL)
Reducer: Sky-Watcher 0.77 Reducer/Flattener
Imaging Camera: ZWO 2600mm / Bin 2x2

Page 83: Owl/ET Cluster (NGC 457)
Color Palette: RGB
Exposures: 20 x 60 sec each filter
Imaging Filters: Antlia V-Pro R/G/B filters
Imaging Telescope: Sky-Watcher 150DX f/6.2 (924mm EFL)
Reducer: Sky-Watcher 0.77 Reducer/Flattener
Imaging Camera: ZWO 2600mm / Bin 2x2

Page 84: M44 Beehive Cluster
Color Palette: RGB
Exposures: 20 x 180 sec each filter
Imaging Filters: Antlia V-Pro R/G/B filters
Imaging Telescope: Sky-Watcher Esprit 80ED f/5 (400mm EFL)
Reducer/Flattener: Standard Sky-Watcher Flattener
Imaging Camera: ZWO 2600mm / Bin 2x2

Page 85: M41 (The Little Beehive)
Color Palette: RGB
Exposures: 16 x 30 sec each filter
Imaging Filters: Antlia V-Pro R/G/B filters
Imaging Telescope: Sky-Watcher 150DX f/6.2 (924mm EFL)
Reducer: Sky-Watcher 0.77 Reducer/Flattener
Imaging Camera: ZWO 2600mm / Bin 2x2

Page 86: M13 (Great Hercules Globular Cluster)
Color Palette: RGB
Exposures: 20 x 15 sec each filter
Imaging Filters: Antlia V-Pro R/G/B filters
Imaging Telescope: Sky-Watcher 150DX f/6.2 (924mm EFL)
Reducer: Sky-Watcher 0.77 Reducer/Flattener
Imaging Camera: ZWO 2600mm / Bin 2x2

Page 87: M92
Color Palette: RGB
Exposures: 60 x 30 sec each filter
Imaging Filters: Antlia V-Pro R/G/B filters
Imaging Telescope: Sky-Watcher 150DX f/6.2 (924mm EFL)
Reducer: Sky-Watcher 0.77 Reducer/Flattener
Imaging Camera: ZWO 2600mm / Bin 2x2

Page 88: M35
Color Palette: RGB
Exposures: 16 x 30 sec each filter
Imaging Filters: Antlia V-Pro R/G/B filters
Imaging Telescope: Sky-Watcher Esprit 80ED f/5 (400mm EFL)
Reducer/Flattener: Standard Sky-Watcher Flattener
Imaging Camera: ZWO 2600mm / Bin 2x2

Page 89: M56
Color Palette: RGB
Exposures: 20 x 60 sec
Imaging Filters: Antlia V-Pro Luminance
Imaging Telescope: Sky-Watcher 150DX f/6.2 (924mm EFL)
Reducer: Sky-Watcher 0.77 Reducer/Flattener
Imaging Camera: ZWO 2600mc / Bin 2x2

Page 90: M50
Color Palette: RGB
Exposures: 91 x 120 sec
Imaging Filters: Antlia V-Pro Luminance
Imaging Telescope: Sky-Watcher Esprit 80ED f/5 (400mm EFL)
Reducer/Flattener: Standard Sky-Watcher Flattener
Imaging Camera: ZWO 2600mc / Bin 2x2

Page 91: Double Cluster (NGC 869 & NGC 884)
Color Palette: RGB
Exposures: 120 x 30 sec
Imaging Filters: Antlia V-Pro Luminance
Imaging Telescope: Sky-Watcher 150DX f/6.2 (924mm EFL)
Reducer: Sky-Watcher 0.77 Reducer/Flattener
Imaging Camera: ZWO 2600mc / Bin 2x2

Page 92: M46 w/NGC 2438
Color Palette: RGB
Exposures: 71 x 120 sec
Imaging Filters: Antlia V-Pro Luminance
Imaging Telescope: Sky-Watcher Esprit 80ED f/5 (400mm EFL)
Reducer/Flattener: Standard Sky-Watcher Flattener
Imaging Camera: ZWO 2600mc / Bin 2x2

Page 93: M47
Color Palette: RGB
Exposures: 71 x 120 sec
Imaging Filters: Antlia V-Pro Luminance
Imaging Telescope: Sky-Watcher Esprit 80ED f/5 (400mm EFL)
Reducer/Flattener: Standard Sky-Watcher Flattener
Imaging Camera: ZWO 2600mc / Bin 2x2

Page 94: M5 (Globular Cluster)
Color Palette: RGB
Exposures: 60 x 30 sec each filter
Imaging Filters: Antlia V-Pro R/G/B filters
Imaging Telescope: Sky-Watcher 150DX f/6.2 (924mm EFL)
Reducer: Sky-Watcher 0.77 Reducer/Flattener
Imaging Camera: ZWO 2600mm / Bin 2x2

Page 95: M45 (Pleiades)
Color Palette: RGB
Exposures: 12 x 180 sec each filter
Imaging Filters: Antlia V-Pro R/G/B filters
Imaging Telescope: Sky-Watcher Esprit 80ED f/5 (400mm EFL)
Reducer/Flattener: Standard Sky-Watcher Flattener
Imaging Camera: ZWO 2600mm / Bin 2x2

Page 96: Coathanger Cluster (Collinder 399)
Color Palette: RGB
Exposures: 60 x 60 sec each filter
Imaging Filters: Antlia V-Pro R/G/B filters
Imaging Telescope: Sky-Watcher Esprit 80ED f/5 (400mm EFL)
Reducer/Flattener: Standard Sky-Watcher Flattener
Imaging Camera: ZWO 2600mm / Bin 2x2

Page 98: Comet C/2023 A3 (Tsuchinshan-ATLAS)
Color Palette: RGB
Exposures: 12 x 4 sec
Imaging Filters: Antlia V-Pro Luminance
Imaging Telescope: Sky-Watcher Esprit 80ED f/5 (400mm EFL)
Reducer/Flattener: Standard Sky-Watcher Flattener
Imaging Camera: ZWO 294mc / Bin 2x2

Page 99: Full Moon
Color Palette: Mono
Exposures: 700 x 0.1ms sec
Imaging Filters: Antlia V-Pro Luminance
Imaging Telescope: Sky-Watcher 150DX f/6.2 (924mm EFL)
Reducer/Flattener: Sky-Watcher 0.77 Reducer/Flattener
Imaging Camera: ZWO 2600mm / Bin 2x2

Page 100: Moon - Mare Tranquillitatis / Serenitatis
Color Palette: Mono
Exposures: 700 x 0.1ms sec
Imaging Filters: Antlia V-Pro Luminance
Imaging Telescope: Sky-Watcher 150DX f/6.2 (924mm EFL)
Reducer/Flattener: Sky-Watcher 0.77 Reducer/Flattener
Imaging Camera: ZWO 2600mm / Bin 2x2

Page 101: Moon - Mare Imbrium
Color Palette: Mono
Exposures: 700 x 0.1ms sec
Imaging Filters: Antlia V-Pro Luminance
Imaging Telescope: Sky-Watcher 150DX f/6.2 (924mm EFL)
Reducer/Flattener: Sky-Watcher 0.77 Reducer/Flattener
Imaging Camera: ZWO 2600mm / Bin 2x2

Page 102: Jupiter with moon shadow
Color Palette: RGB
Exposures: 2000 x 5ms
Imaging Filters: Antlia V-Pro Luminance
Imaging Telescope: Sky-Watcher 180 Maksutov f/15 (2700mm EFL)
Reducer/Flattener: None
Imaging Camera: ZWO 294mc / Bin 2x2

Page 103: Jupiter with moon shadow & Great Red Spot
Color Palette: RGB
Exposures: 2000 x 5ms
Imaging Filters: Antlia V-Pro Luminance
Imaging Telescope: Sky-Watcher 180 Maksutov f/15 (2700mm EFL)
Reducer/Flattener: None
Imaging Camera: ZWO 294mc / Bin 2x2

Page 104: Mars with polar ice cap (white feature)
Color Palette: RGB
Exposures: 13,000 x 10ms
Imaging Filters: Antlia V-Pro Luminance
Imaging Telescope: Sky-Watcher 180 Maksutov f/15 (2700mm EFL)
Reducer/Flattener: None
Imaging Camera: ZWO 294mc / Bin 2x2

Page 105 Saturn with its shadow on the rings
Color Palette: RGB
Exposures: 2000 x 5ms
Imaging Filters: Antlia V-Pro Luminance
Imaging Telescope: Sky-Watcher 180 Maksutov f/15 (2700mm EFL)
Reducer/Flattener: None
Imaging Camera: ZWO 294mc / Bin 2x2

www.ingramcontent.com/pod-product-compliance
Lightning Source LLC
LaVergne TN
LVRC100151061126
851857LV00009B/39

9798997197490